Lecture Notes in Business Information Processing 478

Series Editors

LNBIP reports state-of-the-art results in areas related to business information systems and industrial application software development – timely, at a high level, and in both printed and electronic form.

The type of material published includes

- Proceedings (published in time for the respective event)
- Postproceedings (consisting of thoroughly revised and/or extended final papers)
- Other edited monographs (such as, for example, project reports or invited volumes)
- Tutorials (coherently integrated collections of lectures given at advanced courses, seminars, schools, etc.)
- Award-winning or exceptional theses

LNBIP is abstracted/indexed in DBLP, EI and Scopus. LNBIP volumes are also submitted for the inclusion in ISI Proceedings.

Yu Maemura · Masahide Horita · Liping Fang · Pascale Zaraté
Editors

Group Decision and Negotiation in the Era of Multimodal Interactions

23rd International Conference on Group Decision and Negotiation, GDN 2023 Tokyo, Japan, June 11–15, 2023 Proceedings

Editors
Yu Maemura
University of Tokyo
Tokyo, Japan

Masahide Horita
University of Tokyo
Tokyo, Japan

Liping Fang
Toronto Metropolitan University
Toronto, ON, Canada

Pascale Zaraté
Toulouse University
Toulouse, France

ISSN 1865-1348 ISSN 1865-1356 (electronic)
Lecture Notes in Business Information Processing
ISBN 978-3-031-33779-6 ISBN 978-3-031-33780-2 (eBook)
https://doi.org/10.1007/978-3-031-33780-2

This Springer imprint is published by the registered company Springer Nature Switzerland AG
The registered company address is: Gewerbestrasse 11, 6330 Cham, Switzerland

Preface

The Series of Annual International Conferences on Group Decision and Negotiation (GDN) has provided a stimulating forum for disseminating, discussing, and critiquing the latest research on the theory and practice of group decision and negotiation. GDN conferences have provided invaluable opportunities for participants to share and exchange ideas and have also been a catalyst for collaborative partnerships between scholars who are passionate about discovering fundamental knowledge about group decision-making and negotiations. GDN conferences have taken place every year since 2000 with two exceptions (2000, Glasgow, UK; 2001, La Rochelle, France; 2002, Perth, Australia; 2003, Istanbul, Turkey; 2004, Banff, Canada; 2005, Vienna, Austria; 2006, Karlsruhe, Germany; 2007, Mont Tremblant, Canada; 2008, Coimbra, Portugal; 2009, Toronto, Canada; 2010, Delft, The Netherlands; 2011, Cancelled; 2012, Recife, Brazil; 2013, Stockholm, Sweden; 2014, Toulouse, France; 2015, Warsaw, Poland; 2016, Bellingham, USA; 2017, Stuttgart, Germany; 2018, Nanjing, China; 2019, Loughborough, UK; 2020, Cancelled; 2021, Toronto, Canada (but held online); and 2022, held as a virtual event).

The 23rd International Conference on Group Decision and Negotiation (GDN 2023) was held at the University of Tokyo, Japan, from June 11th to 15th in hybrid format. A total of 102 submissions were received, spanning nine streams related to GDN. After a thorough and careful review process by an international scientific committee, eleven papers were chosen for publication in this volume entitled "**Group Decision and Negotiation in the Era of Multimodal Interactions**". The eleven papers in this volume are organized into three sections, showing the wide spectrum of research that was presented at GDN 2023.

The first section, "**Taking a Step Back: Critically Re-examining Technology Interactions with Group Decision and Negotiation**", presents two papers. Meyer and Schoop provide a valuable and informative systematic literature review for e-negotiation researchers and practitioners. Kaya and Schoop then analyze and evaluate the performance accuracy of predictive machine learning models in response to information growth. Generative artificial intelligence (AI) applications took the world by storm this year, and these first two chapters help remind the GDN community about the importance of systematic and critical research that examines the fundamental linkages between technology and group decision-making and negotiation activities.

The second section of this volume contains four papers related to "**Preference Modeling and Multi-Criteria Decision-Making**". Maia da Silva et al. and Wachowicz et al. provide excellent examples of how preference modeling can be applied to guide investment decisions and contract negotiations. Wachowicz and Czekajski as well as Paulsson and Larsson then outline important theoretical concepts for modeling complex preferences and stakeholder input.

The last section on "**Conflict Modeling and Distributive Mechanisms**" contains theoretical and practical developments towards conflict resolution and distribution problems. Asa et al. and Kato both provide theoretical attempts to explore the concept of permissibility, and highlight the deep roots of contractualism within the GDN community. Similarly, Ziaei and Kilgour reveal important observations on their evaluation of algorithmic allocation procedures against efficiency and fairness criteria. Mirnasl et al. then report thought-provoking implications concerning their application of the graph model for conflict resolution (GMCR) to regulatory conflicts involving multiple levels of government and stakeholders. Finally, Kitadai et al. explore the theoretical effectiveness of a proposed allocation mechanism that weighs rewards to data providers based on their contributions.

The GDN 2023 conference and this published volume represent the efforts and collaboration among a strong, interdependent, and global network of GDN researchers. In particular, we would like to take this opportunity to express our particular appreciation to the Honorary Chair of GDN 2023, Rudolf Vetschera, for his contributions in organizing GDN 2023 and to the Group Decision and Negotiation (GDN) Section, Institute for Operations Research and the Management Sciences (INFORMS). We are also grateful to all the Stream Organizers: Liping Fang, Keith W. Hipel, and D. Marc Kilgour (Conflict Resolution); Femke Bekius and L. Alberto Franco (Design, Use, and Evaluation of GDN Support); Danielle Costa Morais and Tomasz Wachowicz (Preference Modeling for Group Decision and Negotiation); Mareike Schoop, Rudolf Vetschera, and Muhammed-Fatih Kaya (Negotiation Support Systems and Studies (NS3)); Zhen Zhang, Yucheng Dong, Francisco Chiclana, and Enrique Herrera-Viedma (Intelligent Group Decision Making and Consensus Process); Masahide Horita and Hiroyuki Sakakibara (Group Decision and Negotiations for International Project Management); Pascale Zaraté (Collaborative Decision Making Processes); Haiyan Xu, Shawei He, and Shinan Zhao (Risk Evaluation and Negotiation Strategies); and Fuad Aleskerov and Alexey Myachin (Network and Decision Analysis of Political and Group Connections).

Special thanks go to the following reviewers for their informative and prompt reviews of papers: Luciana Alencar, Mischel Carmen Belderrain, Suzana Daher, Gert-Jan de Vreede, Liping Fang, Luis Alberto Franco, Eduarda Frej, Dorota Górecka, Shawei He, Masahide Horita, Muhammed-Fatih Kaya, Alexandre Leoneti, Yu Maemura. Danielle Morais, Hanna Nurmi, Leandro C. Rego, Lucia Reis Peixoto Roselli, Ewa Roszkowska, Maisa Silva, Hao Sun, Takahiro Suzuki, Rudolf Vetschera, Tomasz Wachowicz, Junjie Wang, Shikui Wu, Yi Xiao, Haiyan Xu, Saied Yousefi, Fahimeh Zaiei, and Shinan Zhao.

We are also thankful to the staff at Springer for their excellent support.

April 2023

Yu Maemura
Masahide Horita
Liping Fang
Pascale Zaraté

Organization

Honorary Chair

Rudolf Vetschera	University of Vienna, Austria

General Chairs

Masahide Horita	University of Tokyo, Japan
Liping Fang	Toronto Metropolitan University, Canada

Program Chairs

Yu Maemura	University of Tokyo, Japan
Pascale Zaraté	Université Toulouse 1 Capitole, France

Program Committee

Adiel Teixeira de Almeida	Federal University of Pernambuco, Brazil
Amer Obeidi	University of Waterloo, Canada
Bo Yu	Dalhousie University, Canada
Bogumił Kamiński	Warsaw School of Economics, Poland
Danielle Costa Morais	Universidade Federal de Pernambuco, Brazil
Ewa Roszkowska	Bialystok University of Technology, Poland
Fran Ackermann	Curtin University, Australia
Fuad Aleskerov	HSE University, Russia
Ginger Ke	Memorial University of Newfoundland, Canada
G.-J. de Vreede	University of South Florida, USA
Haiyan Xu	University of Aeronautics and Astronautics, China
Hannu Nurmi	University of Turku, Finland
Jing Ma	Xi'an Jiaotong University, China
John Zeleznikow	La Trobe University, Australia
José María Moreno-Jiménez	University of Zaragoza, Spain
Keith Hipel	University of Waterloo, Canada
Kevin Li	University of Windsor, Canada
Liping Fang	Toronto Metropolitan University, Canada

Luis Dias University of Coimbra, Portugal
Marc Kilgour Wilfrid Laurier University, Canada
Masahide Horita University of Tokyo, Japan
Melvin F. Shakun New York University, USA
Pascale Zaraté Université Toulouse 1 Capitole, France
Przemyslaw Szufel Warsaw School of Economics, Poland
Raimo Hamalainen Aalto University, Finland
ShiKui Wu Lakehead University, Canada
Tomasz Szapiro Warsaw School of Economics, Poland
Tomasz Wachowicz University of Economics in Katowice, Poland
Tung X. Bui University of Hawai'i at Mānoa, USA
Yu Maemura University of Tokyo, Japan
Yufei Yuan McMaster University, Canada

Local Organizing Committee

Takahiro Suzuki University of Tokyo, Japan
Madoka Chosokabe Tottori University, Japan
Masahide Horita University of Tokyo, Japan
Yu Maemura University of Tokyo, Japan
Muhamad Tajammal Khan University of Tokyo, Japan
Shengchi Ma University of Tokyo, Japan

Contents

Taking a Step Back: Critically Re-examining Technology Interactions with Group Decision and Negotiation

Taxonomy of Styles, Strategies, and Tactics in E-Negotiations

Marlene Meyer(✉) and Mareike Schoop

University of Hohenheim, Schwerzstrasse 40, 70599 Stuttgart, Germany
{marlene.meyer,schoop}@uni-hohenheim.de

Abstract. Negotiation are essential in every day's life, increasingly via electronic media. Due to a huge amount of negotiation styles, strategies, and tactics presented in the literature, it is unclear which combinations of strategies and tactics for a concrete style can be applied in (electronic) negotiations. Therefore, we conducted a systematic literature review identifying combinations of negotiation styles, strategies, and tactics in electronic negotiations. The findings were consolidated into a taxonomy and patterns of the combinations were generated.

Keywords: E-Negotiation · Systematic Literature Review · Negotiation Styles · Negotiation Strategies · Negotiation Tactics · Taxonomy

1 Introduction and Background

Negotiation in general are communication and decision-making processes between at least to parties who exchange arguments with the aim to solve conflictual situation [35, 36, 64]. "Conflict is at the center of negotiations. Conflict in terms of negotiation does not necessarily have to be an actual dispute or disagreement. It suffices that there is insufficient information about the other's wishes or about the other's perception of the events in order to create a conflict." [53].

The Thomas-Kilmann-Conflict Mode Instrument (TKI) identifies a personal conflict style of individuals that can be applied to handle conflictual situations [71], such as negotiation [26]. TKI considers two dimensions how an individual can behave in conflict situations: (1) assertiveness, i.e. the extent to which an individual is concerned with their own interests; (2) cooperativeness, i.e. the extent to which an individual is concerned with the partner's interests [71, 72]. Further, TKI can be applied to identify the partner's style in negotiations [25, 83]. Based on the two dimensions, TKI defines the five styles competing, accommodating, avoiding, collaborating, and compromising. A *competing* negotiator is only concerned with their own interests no matter how it affects the relationship to the partners. An *accommodating* negotiator is concerned with the relationship to the partner, i.e., the negotiator is yielding to the partner's view. An *avoiding* negotiator is neither concerned with their own interests nor with those of the partner, i.e. the negotiator postpones an issue or eludes the negotiation. A *compromising* negotiator is concerned with both their own interests and with those of the partner and tries to find a

Y. Maemura et al. (Eds.): GDN 2023, LNBIP 478, pp. 3–19, 2023.
https://doi.org/10.1007/978-3-031-33780-2_1

conclusion that partially satisfy all parties, e.g., by splitting the difference or exchanging concessions. A *collaborating* negotiator is concerned with their own interests and those of the partner and tries to find a solution that satisfies all parties [71, 72]. In our research, the TKI styles provide a basis for different types of negotiation partners and are defined as negotiation styles.

For allocating findings in the literature to negotiation styles, strategies, and tactics, they have to be defined. Negotiation styles are orientations how someone can negotiate and are dependent on the importance of the outcome and the relationship [46]. Negotiation strategies consider an overall plan of the negotiation according to selection and re-evaluation of priorities and used negotiation styles [27, 43]; thus some negotiation strategies are more applicable to a particular style than other strategies. Negotiation tactics are techniques to apply specific negotiation styles and strategies [27, 43] and thus vary during the negotiation process.

To choose the appropriate strategies, and tactics seems to be a difficult task especially for novice negotiators [52]. The definitions of strategies and tactics vary in the literature. Some findings have different terms for identical strategies or tactics, e.g. large concession vs. being soft [9, 74, 77]. In other findings, identically described strategies or tactics are allocated in some cases to strategies and in other cases to tactics, e.g. Tit-for-Tat [74, 86]. Thus, the aim of this research contribution is to classify styles, strategies, and tactics, and group similar strategies/ tactics with consistent naming and consistent allocation to a strategy or tactic. Whether they are allocated to strategies or tactics depends on prior definitions. The following research question is formulated:

RQ1. To what extent negotiation styles, strategies, and tactics in e-negotiation can be classified?

Electronic negotiations (e-negotiations) comprise electronic argument exchange from simple offers to complex communication and thus can be used to apply different negotiation models, such as negotiation agents (as negotiation partner) and negotiation support (supporting human negotiators in complex negotiations) [10, 65].

Negotiation support systems (NSSs) are communication and information systems, which support the negotiations process by gathering information, structuring, and generating alternatives [10, 65].

Once negotiation styles, strategies, and tactics are classified, we will investigate whether specific negotiation styles, strategies, and tactics are mentioned together in the literature and can be linked together as a pattern. Each pattern contains one style of the TKI, and the matching strategies and tactics found in the literature. The aim of the patterns is to possess different combinations of strategies and tactics for a concrete style to apply in e-negotiations. Our investigation will focus on decision support in bilateral semi-structured human-machine negotiations. To the best of our knowledge, an extensive model of the mentioned aim is missing and thus the following research question is formulated:

RQ2. Which patterns can be identified between the classified negotiation styles, strategies, and tactics?

2 Methodological Approach

This chapter describes how the research was conducted. First, negotiation styles, strategies, and tactics were identified through a systematic literature review. Second, based on the findings of the literature review a taxonomy was created to determine a classification of the identified negotiation styles, strategies, and tactics (RQ1) and finally patterns of the classified styles, strategies, and tactics in e-negotiations were generated (RQ2).

2.1 Systematic Literature Review

The systematic literature review [81, 82] is shown in Table 1. Google scholar was chosen as search engine due to its broad coverage of major databases and publishers. The aim of this literature review was to identify styles, strategies, and tactics in e-negotiation in general and in the application of a software agent. Culture is not explicit considered in this research.

Table 1. Steps of Systematic Literature Review

Step	Description	Remaining Papers
Search	Google scholar as search engine; Search strings: 1) "negotiation style" OR "negotiation strategy" OR "negotiation tactic" "electronic negotiation" -culture 2) "negotiation style" OR "negotiation strategy" OR "negotiation tactic" "e-negotiation" -culture 3) "negotiation style" OR "negotiation strategy" OR "negotiation tactic" "software agent" "e-negotiation" -culture"	876
Exclusion 1	Reason for exclusion: duplicates, language (not English), no access	397
Exclusion 2	Reason for exclusion: missing relevance of title and abstract	167
Exclusion 3	Reason for exclusion: missing relevance of the full text	139

In total 876 findings were identified within three search algorithms. However, these findings comprise various duplicates, missing access on full paper, and non-English publications, which had to be excluded. In Exclusion 2, the relevance was examined according to the title and abstract of the publications. After the third exclusion, which examined the relevance of full text, 139 findings were identified as relevant.

2.2 Taxonomy Development

A taxonomy according to Nickerson [55] was created to classify negotiation styles, strategies, and tactics.

Firstly, a meta-characteristic as "the most comprehensive characteristic that will serve as the basis for the choice of characteristics in the taxonomy" [55] was defined, i.e., negotiation styles, strategies, and tactics in semi-structured bilateral e-negotiation with decision-making focus. Whether negotiation styles, strategies, and tactics were mentioned together in the findings was insignificant for the creation of the categories in the taxonomy.

Afterwards for defining the termination, the subjective and objective ending conditions were characterised. The objective ending conditions were concise, robust, extensible, and explanatory [55].

Finally, the approach of the iterations to create the taxonomy had to be chosen out of inductive (empirical-to-conceptual) or deductive (conceptual-to-empirical) approaches [55]. In the first iteration, we applied an inductive approach based on an unstructured literature search in the field of negotiation styles, strategies, and tactics by analysing 50 publications. Based on our defined meta-characteristic, we identified two dimensions with 19 subdimensions [23, 24, 43, 73, 84]. The second iteration applied an inductive approach based on a systematic literature review. Due to exponentially increasing categories in the dimensions, we had to restructure the taxonomy into three dimensions with nine subdimensions. Now, the subjective ending conditions were met. Due to the restructuring process and remaining literature from the literature review, a third iteration was conducted by applying an inductive approach. After performing the third iteration all objective ending conditions were met [55], especially since all objects were examined.

3 Taxonomy

The taxonomy (see Fig. 1) is characterised by three dimensions – styles, meta-strategies, and meta-tactics. The style dimension includes the categories avoiding, accommodating, competitive, collaborative, and compromising which are based on TKI [71, 72] and thus verified by the literature. Meta-strategies and meta-tactics contain further subdimensions.

Meta-strategies. The category meta-strategies provides strategies and contains the subdimensions central concern, concession-making approach, concession-making degree, information exchange, strategy composition, and other approaches; which are evaluated in the following according to the findings in the literature review.

Central concern as a subdimension contains the basic orientation of how negotiators are concern with their own interests and/or their partner's interests. The TKI model describes assertiveness as being concernd with one' own behaviour and cooperativeness as being concerned with the partner's interest. Being only focused on own interests, an individual is denoted as competing. Negotiators who are only interested in their own concern, only behave according to their goals, do not consider knowledge about the partner or employ a "take-it-or-leave-it" strategy [6, 8, 9, 12, 86]. Negotiators who are only interest in the partner's concern, try to achieve an outcome that suits the partner's needs, e.g., by employing the strategy to ask the partner to make an offer [12]. Negotiators who are neither interested in their own concern nor in their partner's concern, e.g. postpone or cancel the negotiation [71, 72]. Negotiators who are interested in their own and partner's concern, can be collaborating or compromising negotiators; compromising negotiators split the difference and thus find a partially satisfied conclusion for all parties; collaborative negotators expand the pie of the negotiation by finding a solution that satisfies all parties [71, 72].

	Dimension	Categories				
	Styles	Avoiding	Accommodating	Competitive	Collaborative	Compromising
Meta-strategies	Central concern	Partner's concern	Own concern			
Meta-strategies	Concession-making degree	Large concession	Gradual concession	No concession		
Meta-strategies	Concession-making approach	Random	Trade-offs	Fixed concession strategy	Repetitive strategy	
Meta-strategies	Information exchange	Share information	Withholding information			
Meta-strategies	Strategy composition	Pure strategy	Mixed strategy			
Meta-strategies	Other approaches	Coordination strategy	Relation-based negotiation strategy	MESOArgN	...	
Meta-tactics	Time-dependent	Delayed concession	Immediate concession			
Meta-tactics	Behaviour-dependent	Concession reciprocity	Estimating partner			
Meta-tactics	All	Heuristic models	Mathematical models	Game theory algorithm	Degree of concession	...

Fig. 1. Taxonomy

Solving conflictual situations to achieve an agreement, concessions have to be made by the negotiators [63]. Based on this assumption the subdimensions concession-making degree and concession-making approach were generated.

Concession-making degree defines to what extent a concession is made, namely no, gradual, or large concession. The category in which negotiators do not make any concession [60, 86] or do not want to participate in the negotiation is defined as no concession. Examples of concrete strategies are avoiding discussions [60, 61, 86] or leaving the negotiation [12]. Gradual concession describes more/ fewer concessions being made depending on the concrete strategies. For example, an optimistic opening offer is provided, followed by small concessions [47]. More precisely, the negotiator is willing to make further large concessions if the partner will also make large concessions, or the negotiator tries to split the difference [12]. Making large concessions could have multiple reasons. Some negotiators might want to conclude the negotiation as quickly as possible [18, 86]; for others achieving an agreement at all is more relevant than no agreement at the end of a negotiation [22, 37, 56, 76]. With these three categories, all findings of the literature review are included.

The **concession-making approach** comprises how an individual wants to proceed during making concession, e.g., willingness to trade off [37, 87], trial-and-error or random [31, 47], and thus is relevant to determine the own strategy in a negotiation. This subdimension contains the categories random, trade-offs, fixed-concession strategy, and repetitive strategies.

A random approach possesses a set of feasible issue combinations and picks random a combination of issues as an offer, which leads to unpredictable offers during the negotiation [6, 8, 22, 32]. In trade-offs, issues of low importance are traded for issues of high importance for oneself, i.e., an individual make high concessions on issues with low importance and low concessions on issues with high importance [6, 19, 32, 37, 42]. In fixed concession strategies, the concession making depends on the urgency when the negotiation should end or on the necessity, an outcome is needed. A greater urgency causes greater concessions [42]. The category "Repetitive strategies" includes routine-based strategies, such as case-based reasoning [50].

Knowing, e.g., the intention of the partner enables a negotiator to influence the partner's behaviour to its own benefits [35]. Thus, **exchanging information** with the partner is a part of negotiation as well as of the taxonomy. Dependent on negotiator's goal the extent of information exchanged, and the extent of trustworthiness information vary, i.e., one wants to hide information or share them with the partner.

In "share information", negotiators are willing to exchange information about themselves and their preferences to the partner [18]. The shared information can be truthful, wrong, or irrelevant [76], e.g. claiming an issue to be important whilst it is really not important for the negotiator. Further information can be shared with an intentional use [9], i.e., the negotiator only shares information by influencing the beliefs and intentions of the partner [60]. By withholding information, the negotiator intends to obfuscate its truth intentions and beliefs, e.g. contradictory offers. Mostly this strategy is applied in distributive negotiations [9, 76].

The approach to achieve goals in the negotiation (**strategy composition**) might change through receiving new information about the negotiation setting or the partner, and thus the applied strategies might have to be changed too. If one strategy is applied for all issues or over time, it is called pure strategy [88]. If multiple strategies are applied for different issues or over time, it is called mixed strategy. Based on our literature review, especially agents are applying negotiation strategies, which are changing over time and thus apply a mixture of strategies [5, 14, 38, 80, 88].

The subdimension **other approaches** consolidates further findings of the literature review [19, 22, 56, 66, 68], which are only mentioned once or twice and are concrete applications. Thus, these findings are unclassifiable to the already defined subdimensions, such as coordination strategy, relation-based negotiation strategy or MESOArgN. As there are no limitations of future findings of concrete applications one category is defined as "…" to consolidate further findings for strategies, which cannot be classified in the remaining categories of the taxonomy.

Meta-tactics. Time-dependent, resource-dependent, and behaviour-dependent are common terms for negotiation tactics in negotiation literature [2, 16, 17, 58, 77]. Due time-dependent, resource-dependent, and behaviour-dependent describe orientations of tactics, they were considered as subdimensions in the dimension meta-tactics. Behaviour-dependent tactics are based on the negotiation partner's attitude, e.g., tit-for-tat. In time-dependent tactics the negotiator's attitude changes over time, e.g., near deadline a negotiator changed their tactics from tough to conceding. Resource-dependent tactics contain changes in resources, i.e., similar to time-dependent, but the concession pattern can differ [23]. Tactics belonging to a combination of meta-tactics can be categorised to combined meta-tactics. After applying the literature on our taxonomy, the subdimensions time-dependent, behaviour-dependent and all (time-dependent, resource-dependent, and behaviour dependent) were identified.

Tactics affecting different time points in a negotiation [87] are categories of the subdimension **time-dependent**, e.g. when concessions are made. The two defined categories are delayed concession and immediate concession.

In delayed concessions, first concessions are made in later periods, i.e., starting tough and change to making (large) concession while approaching the end of the deadline, e.g., Boulware tactic [4, 13, 54, 67, 76]. Immediate concessions are characterised by making concession at the beginning of a negotiation, i.e., in the first offer. Concessions can remain monotonic or become smaller to no concessions over time, e.g., conceder tactic [4, 18, 67, 76, 86].

Behaviour-dependent tactics contain tactics how a negotiator behaves, e.g., the extent of willingness to make concessions if receiving concessions (Concession reciprocity) or assessing the partner to gain knowledge (Estimating partner).

Concession reciprocity describes the tendency of a negotiator to reciprocate concession, e.g., tit-for-tat [8, 31, 41, 47, 79]. Estimating the partner's preferences can enable a sufficient negotiation outcome for oneself. For example, the reservation value or the probability that the partner will accept a certain offer can be estimated [8]. Further behaviour-dependent tactics (e.g., power) affect multiple categories and are dependent on the partner, and thus are consolidated in an overall behaviour.

The subdimension "**all**" comprises time-dependent, resource-dependent, and behaviour dependent. However, it does not imply that all three meta-tactics perform together, rather than the categories can be time, behaviour, and resource related. It mainly contains models, functions, and algorithms, i.e., concrete applications of tactics, which can be applied as a negotiation partner in e-negotiations, and includes the categories heuristic models, mathematical models, game theoretic models, degree of concession, and "…". The category "…" contains further findings of the literature review, which are concrete application and unclassifiable into the previous defined categories, e.g., BOA, KDE, Zeuthen strategy, and QO-Agent [54, 60, 79]. This category enables the taxonomy to be extendable for further investigations or new findings.

Heuristic models search the scope of the negotiation incompletely, i.e., the negotiation outcome can be suboptimal [33], but more realistic [15, 23, 61], i.e., these models should be investigated further as negotiation partner in electronic human-machine negotiations. Heuristic models comprise several models, such as, artificial neural networks, rule-based algorithms, argumentation-based algorithm [18, 21, 33, 57, 61], genetic algorithms, Q-Learning [38, 50], fuzzy algorithms [41, 49], reinforcement learning [11, 31], and regression [11]. Mathematical models comprise especially mathematical functions (e.g., linear, exponential, or logarithmic) [20, 34, 38, 48, 76] and algorithms modelling tactics [7, 41, 42, 67, 69] and thus outline how concessions are changed over altered behaviour, remaining time, or remaining resource. These models can be applied as support or negotiation partner in electronic human-machine-negotiations. Game theoretic models include models based on game theory approach, such as Rubinstein, Markov chains, and Bayesian learning [8, 87]. Agents using game theoretic models are defined as completely rational [33]. Due to rationality of such models, they tend to be not applicable as negotiation partner in e-negotiations. Whereas applying game theoretic models, e.g. identifying strategy changes by using Markov chains or Bayesian learning [8], can support the negotiators in a human negotiation. "Degree of concession" depicts on a precise level to what extent a concession is made within single issues of an offer (no, small, moderate, large, complete) [7, 30, 70, 74, 75]. The selections cover all relevant degrees of concession in our taxonomy. The selection of a model relies on the framing of a concrete negotiation and has to be re-examined for each negotiation.

The described taxonomy enables to show the behaviour of a negotiator without being influenced by the negotiation partner or the setting of the negotiation. There are further categories, which are influenced by the partner or the setting, such as power, anchoring, issue consideration with multi-issues offers and single issue offers. These categories are classified as an overall behaviour in this research contribution. Power categorises the equality of relationship between the negotiation partners. If the power between the parties is equal, the behaviour of the negotiation is relationship oriented, e.g., accommodating or collaborative. If the power between the parties is unequal, the negotiator's behaviour can be self-centric, e.g., competitive or aggressive, in the negotiation [9, 58, 85]. The first offer serves as a landmark for the negotiation parties in the remaining negotiation and is called anchoring [1]. Issue consideration determines the number of issues considered in a single offer, i.e., either single-issue offers, or multi-issues offers. If multiple issues are discussed in the negotiation, single issue offers can be discussed at the beginning of the negotiation, however during the negotiation process the negotiation parties have to switch to multi-issues offers to conclude the negotiation [8, 79].

4 Patterns of Styles

Based on the conceptualised taxonomy and the findings in the literature review, combinations between negotiation styles and strategies and tactics were identified and consolidated into patterns. In this research contribution, patterns are defined as occurrences of the categories in the taxonomy identified in the literature review. A pattern contains all categories mentioned in the literature together with the examined style; regardless how often the categories were mentioned together with the examined style in the literature review. Some patterns have only a few categories which were explicitly mentioned in the literature review, e.g., with the styles accommodating and avoiding. These styles are in general not suitable for a successful negotiation outcome and thus are only mentioned a few times in the findings of the literature review.

Each pattern contains one of the five styles and their identified strategies and tactics (see Fig. 2). Whether negotiation styles, strategies, and tactics were mentioned together is only considered in the patterns and not in the creation of the taxonomy. Thus, some categories in the taxonomy could not be explicitly identified in the patterns. Due to the generality of the taxonomy, i.e., the taxonomy could also be applied to choose tactics for concrete strategies without considering the negotiation style; it is valid, that some categories are not comprised in our considered patterns.

Due to some opposite approaches of strategies or tactics, not all (but some) categories identified in a pattern cannot be applied together. In fact, applying concrete strategies

	Style based patterns	Avoiding	Accommodating	Competitive	Collaborative	Compromising
Meta-strategies	Central concern	-	Partner's concern*	Own concern	Partner's concern, Own concern	Partner's concern, Own concern*
Meta-strategies	Concession-making degree	Large concession, No concession	Large concession	No concession	Large concession, Gradual concession, No concession*	Gradual concession
Meta-strategies	Concession-making approach	Random, Trade-offs, Fixed concession strategy, Repetitive strategy*	Random, Trade-offs, Fixed concession strategy, Repetitive strategy*	Trade-offs	Trade-offs	Random, Trade-offs, Fixed concession strategy, Repetitive strategy*
Meta-strategies	Information exchange	Share information, Withholding information*	Share information, Withholding information*	Share information, Withholding information	Share information	Share information
Meta-strategies	Strategy composition	Pure strategy, Mixed strategy*	Pure strategy, Mixed strategy*	Mixed strategy	Mixed strategy	Pure strategy, Mixed strategy*
Meta-strategies	Other approaches	Coordination strategy, Relation-based negotiation strategy, MESOArgN, ...*	Coordination strategy, Relation-based negotiation strategy, MESOArgN, ...*	Coordination strategy, Relation-based negotiation strategy, MESOArgN, ...*	Relation-based negotiation strategy, MESOArgN	Coordination strategy, Relation-based negotiation strategy, MESOArgN, ...*
Meta-tactics	Time-dependent	Delayed concession, Immediate concession*	Immediate concession	Delayed concession, Immediate concession	Delayed concession, Immediate concession	Delayed concession, Immediate concession
Meta-tactics	Behaviour-dependent	Concession reciprocity, Estimating partner*	Concession reciprocity, Estimating partner*	Concession reciprocity, Estimating partner*	Concession reciprocity	Concession reciprocity, Estimating partner*
Meta-tactics	All	Heuristic models, Mathematical models, Game theory algorithm, Degree of concession, ...*	Mathematical models	Heuristic models, Mathematical models, Degree of concession	Heuristic models, Mathematical models, Degree of concession, ...	Mathematical models

Fig. 2. Identified patterns

and tactics in a negotiation, reasonable strategy-tactic-combinations have to be utilised, which are explained in the following.

In the pattern **avoiding**, negotiators neither concern about their own interest nor about their partner's interest (subdimension central concern) [3, 86] and either make large concession [61] or no concession [60, 61, 86] (subdimension concession-making degree). Within the style avoiding, large concessions tend to be made when the negotiator wants to end the negotiation as fast as possible. No concessions tend to be made when the negotiator does not want to participate at all and try to postpone the negotiation. Both characteristics are accompanied by the definition of avoiding.

The pattern **accommodating** contains large concession (subdimension concession-making degree) [18, 86], immediate concession (subdimension time-dependent) [18, 86], and linear function in the mathematical models (subdimension all) [48]. Align to the style definition, the accommodating pattern is characterised by quick, large concession [18, 59], and immediate concession [18, 59] as identified categories. Linear function (category mathematical models) [48] as implementation of the concession matches the category immediate concession.

The **competitive** pattern comprises the following seven categories in five subdimensions of meta-strategies: own concern [6, 9, 86] and partner's concern [87] (subdimension central concern); no concession (subdimension concession-making degree) [86]; trade-offs (subdimension concession-making approach) [86]; share information (no implication of truthful information) [9] and withholding information [9, 76] in the subdimension information exchange; and mixed strategy (subdimension strategy composition) [4, 13, 54, 67, 76].

In "central concern", concerning about the own interests was mentioned several times in the literature review, which suits the definition of competitive style. The subdimension information exchange in the competitive patterns includes the categories share information [9] and withholding information [9, 76]. Sharing information in this context refers to considered information exchange, which suits the competitive negotiation style. The following five categories were identified in meta-tactics: delayed [4, 13, 54, 67, 69, 76, 77] and immediate [4, 13, 54, 67, 69, 75, 76] concession in the time-dependent subdimension; heuristic models [87], mathematical models [42, 48, 54, 67, 69], and degree of concessions [7, 9, 18, 75, 76] in the "all" subdimension. In our findings, immediate concession in the time-dependent subdimension appears always together with delayed concession and the category mixed concession whereas delayed concession [76, 77] was also mentioned standalone. Thus, immediate concession only has a small impact on the style, which is supported by the style definition. However, it can influence the negotiator to be less tough over time. Trying to make no concession (subdimension concession-making degree) and trading less important issues for important issues for themselves (trade-offs in concession-making approach subdimension) represent the competitive style as only concerning about the own interests.

The **collaborative** pattern comprises five subdimension and six categories in the meta-strategies, namely own concern [3] and partner's concern [19, 86] (central concern subdimension), trade-offs [19, 41, 47, 79, 86] (subdimension concession-making approach), share information [18, 76] (information exchange subdimension), mixed strategy [47] in subdimension strategy composition, and Relation-based negotiation strategy and MESOArgN [19, 57] in the subdimension other approaches.

Only considering the own concern was mentioned once without the extent of assertiveness; concerning about the own and partner's interests was mentioned several times in the literature review, which suits the definition of the collaborative style. Sharing information in the collaborative pattern comprises truthful exchange of information. Truthful information exchange enables the negotiator to recognise interests and preferences of the partner and thus enables to exchange highly important issues with low important issues for themselves and their partner (trade-offs). In meta-tactics seven categories in three subdimensions were identified, namely delayed [42], and immediate [47, 69, 86] concession (time-dependent subdimension), concession reciprocity [9, 19, 47, 79] (behaviour-dependent subdimension), and heuristic models [40], mathematical models [41, 42, 69], degree of concession [7, 9, 19, 47, 79], and"..." [54] in the "all" subdimension. In the collaborative pattern delayed concession were mentioned once together with the immediate concession, i.e., depending on the concession reciprocity of the partner a mixed strategy can be applied to counteract on a partner with a less collaborative behaviour.

The pattern **compromising** contains two subdimensions in meta-strategies and two subdimensions in meta-tactics. The following categories were identified in meta-strategies: gradual concession [18] in the concession making degree, and wrong/ irrelevant information are shared in the information exchange [76] subdimension. In meta-tactics immediate concession [75, 76] in the time-dependent subdimension, and mathematical models [48, 76] in the "all" subdimension were identified.

A compromising negotiator is defined as someone with an overall medium interest in their own concerns and in their partner's concerns which can be characterised by self-centric strategies/ tactics (e.g. exchange irrelevant information [3, 76] and delayed concession [18]) in some aspects and partner-focused strategies/tactics (e.g. immediate concession [3, 76]) in other aspects. The pattern describes negotiators focusing on their own concern and simultaneously making concessions towards the partner to achieve the set goals.

Besides the mentioned patterns, two combinations of styles – competitive-compromising and collaborative-competitive – were identified in the literature.

The style dimension **competitive-compromising** comprises mixed strategy [75, 76] (subdimension strategy composition). In meta-tactics immediate [75], and delayed [48] concessions in time-dependent subdimension, and degree of concession [75] in "all" subdimension were identified. Comparing the competitive-compromising pattern with the pattern competitive and the pattern compromising shows that this pattern accompanies the competitive pattern completely and the compromising pattern partially.

The style dimension **collaborative-competitive** contains mixed strategy [4, 13] (subdimension strategy composition). In meta-tactics immediate [4, 13], and delayed [4, 13] concessions in the time-dependent subdimension, and mathematical models [4] in the "all" subdimension were identified. Comparing the collaborative-competitive pattern with the pattern competitive and the pattern collaborative shows that this pattern can be found completely in the competitive pattern and in the collaborative pattern. Both style combinations – competitive-compromising and collaborative-competitive – underline the usefulness of the taxonomy as similar patterns are accompanied by the same category selection.

Further meta-strategies and meta-tactics without specifying a concrete style were mentioned together in the literature review. Due to the focus on patterns related to the negotiation styles, these relations were not discussed further.

5 Outlook

In this research contribution, a literature review was conducted to 1) generate a taxonomy including relations between negotiation styles, strategies, and tactics, and 2) identify patterns of applied negotiation strategies and tactics for a concrete negotiation style. The generated taxonomy includes three dimensions – styles, meta-strategies, and meta-tactics – with nine subdimensions. Five patterns align to the TKI styles and two combinations of styles – compromising-competitive and collaborative-competitive – were identified in the literature review. Individual conflict styles consist of different combinations of TKI styles. Thus, additional investigations could identify further patterns by investigating several combinations of conflict styles. In order to evaluate the taxonomy and the patterns data can be utilised from past negotiation experiments to identify behaviour of negotiators and decode into our taxonomy and identified patterns.

As every research contribution, this research has limitations. The results are limited by the applied search criteria and date of search. However, the generated taxonomy is editable and thus, can be adjusted by further findings. Further, the taxonomy focuses only on strategies and tactics from the decision support context. We are aware that in e-negotiations with humans not only the outcome but also message exchange during offers is relevant and should be investigated in further research. For building patterns, we focused on findings, which were mentioned together with the examined styles, i.e., combinations of meta-strategies and meta-tactics without concrete styles were not considered. Future research could investigate these combinations and examine whether they are related to the already identified patterns. Already presented agents or algorithms from other researcher were further identified in this literature review and summarised in the taxonomy into the categories specific applications, heuristic models, mathematical models, game theory algorithms, and other approaches. As next step, the applicability of the identified agents and algorithms has to be investigated and own algorithms might have to be designed to fully match the identified patterns.

Due to this taxonomy strategic behaviours of negotiators are classified which enables comparison of various behaviours or a self-reflection for a negotiator as well as the possibility to integrate the strategic behaviour into a computational system.

Improving negotiation outcome, a training can be applied [44, 62] in NSSs, e.g. Inspire [39, 78] or Negoisst [51]. Negotiation training should emphasise which negotiation strategies, and tactics are available and offer training on all components. Some researchers indicate that automated agents as negotiation partners can increase training output for novices better than human partners [28, 45]; agents can provide human negotiators with the opportunity to try out negotiation situations with various negotiation styles, strategies and tactics whilest not being distracted from reasoning [29].

For agents to react to human non-deterministic behaviour in human-machine negotiation, agents have to combine various strategies and tactics dynamically during the negotiation process [13]. Furthermore, agents have to possess a pool of negotiation strategies and tactics to initiate a learning process for different types of human negotiators [52]. The identified patterns in this research contribution can generate a foundation for a negotiation agent negotiating dynamically with various conflict styles to train individuals.

References

1. Agrawal, M., Chari, K.: Learning negotiation support systems in competitive negotiations: a study of negotiation behaviors and system impacts. In: 5th International Journal of Intelligent Information Technologies, pp. 1–23 (2009). https://papers.ssrn.com/sol3/papers.cfm?abstract_id=904619
2. Al-Agtash, S.Y., Al-Fahoum, A.A.: An evolutionary computation approach to electricity trade negotiation. Adv. Eng. Softw. **36**(3), 173–179 (2005). https://doi.org/10.1016/j.advengsoft.2004.07.008
3. Al-Jaljouli, R., Abawajy, J., Hassan, M.M., Alelaiwi, A.: Secure multi-attribute one-to-many bilateral negotiation framework for e-commerce. IEEE Trans. Serv. Comput. **11**(2), 415–429 (2018). https://doi.org/10.1109/TSC.2016.2560160
4. Awasthi, S.K., Vij, S., Mukhopadhyay, D., Agrawal, A.J.: Multi Strategy Selection in E-Negotiation. In: Second International Conference on Information and Communication Technology for Competitive Strategies, pp. 1–5. ACM, New York, NY, USA (2016). https://doi.org/10.1145/2905055.2905138
5. Awasthi, S.K., Vij, S.R., Mukhopadhyay, D., Agrawal, A.J.: Multi-strategy Based Automated Negotiation. BGP Based Architecture. In: International Conference on Computing, Communication and Automation, pp. 588–593. IEEE (2016). https://doi.org/10.1109/CCAA.2016.7813789

6. Baarslag, T.: Exploring the Strategy Space of Negotiating Agents. Springer International Publishing, Cham (2016). https://doi.org/10.1007/978-3-319-28243-5
7. Baarslag, T., Gerding, E.H., Aydogan, R., Schraefel, M.C.: Optimal negotiation decision functions in time-sensitive domains. In: International Conference on Web Intelligence and Intelligent Agent Technology, pp. 190–197. IEEE (2015). https://doi.org/10.1109/WI-IAT.2015.161
8. Baarslag, T., Hendrikx, M.J.C., Hindriks, K.V., Jonker, C.M.: Learning about the opponent in automated bilateral negotiation: a comprehensive survey of opponent modeling techniques. Auton. Agent. Multi-Agent Syst. **30**(5), 849–898 (2015). https://doi.org/10.1007/s10458-015-9309-1
9. Baytalskaya, N.: The Effects of Machiavellianism, Perspective Taking, and Emotional Intelligence Components on Negotiation Strategies and Outcomes. Master Thesis, Pennsylvania State University (2017). https://etda.libraries.psu.edu/catalog/8158
10. Bichler, M., Kersten, G., Strecker, S.: Towards a structured design of electronic negotiations. Group Decis. Negot. **12**(4), 311–335 (2003). https://doi.org/10.1023/A:1024867820235
11. Brzostowski, J.W.: Predictive Decision-making Mechanisms Based on Off-line and On-line Reasoning. Dissertation, Swinburne University of Technology (2007)
12. Cairo, O., Olarte, J.G., Rivera-Illingworth, F.: A Negotiation Strategy for Electronic Trade Using Intelligent Agents. In: IEEE/WIC International Conference on Web Intelligence, pp. 194–200. IEEE Comput. Soc (2003). https://doi.org/10.1109/WI.2003.1241193
13. Cao, M., Peng, L.: Enabling computer to negotiate with human in e-commerce: a strategy model. In: 49th Hawaii International Conference on System Sciences, pp. 369–376. IEEE (2016). https://doi.org/10.1109/HICSS.2016.52
14. Cao, M., Luo, X., Luo, X., et al.: Automated negotiation for e-commerce decision making: a goal deliberated agent architecture for multi-strategy selection. Decis. Support Syst. **73**(1), 1–14 (2015). https://doi.org/10.1016/j.dss.2015.02.012
15. Cao, M., Wang, G.A., Kiang, M.Y.: Modeling and prediction of human negotiation behavior in human-computer negotiation. Electron. Commer. Res. Appl. **50**(4), 101099 (2021). https://doi.org/10.1016/j.elerap.2021.101099
16. Carbonneau, R., Vahidov, R.: A multi-attribute bidding strategy for a single-attribute auction marketplace. Expert Syst. Appl. **43**, 42–50 (2016). https://doi.org/10.1016/j.eswa.2015.08.039
17. Carbonneau, R.A., Vahidov, R.M.: A utility concession curve data fitting model for quantitative analysis of negotiation styles. Expert Syst. Appl. **41**(9), 4035–4042 (2014). https://doi.org/10.1016/j.eswa.2013.12.029
18. Chen, E.: An Electronic Market for Agent-supported Business Negotiations. Master thesis, Concordia University (2003). https://spectrum.library.concordia.ca/id/eprint/2420/
19. Cheng, W.-K., Ch, H.-Y.: Conflict resolution in resource federation with intelligent agent negotiation. In: Xu, H. (ed.) Practical Applications of Agent-Based Technology. InTech (2012). https://doi.org/10.5772/36189
20. Choi, H.R., Kim, H.S., Hong, S., Park, Y.J., Park, Y.-S.: Study on the automated negotiation methodology for solving multi attribute negotiation. In: PACIS **63** (2005)
21. El-Fangary, L.M., Kholeif, S., Afify, G.M.: Automated Negotiation Decision Making Approaches: Experimental Study (2008)
22. Etukudor, C., Couraud, B., Robu, V., Früh, W.-G., Flynn, D., Okereke, C.: Automated negotiation for peer-to-peer electricity trading in local energy markets. Energies **13**(4), 920 (2020). https://doi.org/10.3390/en13040920
23. Faratin, P., Sierra, C., Jennings, N.R.: Negotiation decision functions for autonomous agents. Robot. Auton. Syst. **24**(3–4), 159–182 (1998). https://doi.org/10.1016/S0921-8890(98)00029-3

24. Faratin, P., Sierra, C., Jennings, N.R.: Using Similarity criteria to make issue trade-offs in automated negotiations. Artif. Intell. **142**(2), 205–237 (2002). https://doi.org/10.1016/S0004-3702(02)00290-4
25. Fujita, K.: Automated strategy adaptation for multi-times bilateral closed negotiations. In: Ana, B., Michael, H., Alessio, L., Paul, S. (eds.) Proceedings of International Conference on Autonomous Agents and Multi-agent Systems. International Foundation for Autonomous Agents and Multiagent Systems (ACM Digital Library), pp. 1509–1510, Richland, SC (2014). https://dl.acm.org/doi/abs/10.5555/2615731.2616036#d18537779e1
26. Ganesan, S.: Negotiation strategies and the nature of channel relationships. J. Mark. Res. 30(2), p. 183 (1993).https://doi.org/10.2307/3172827
27. Goldman, A.: Negotiation. Theory and Practice. Wolters Kluwer Law Inter-national, Alphen aan den Rijn (2003)
28. Gratch, J,. DeVault, D., Lucas, G.: The benefits of virtual humans for teaching negotiation. In: Traum, D., Swartout, W., Khooshabeh, P. et al. (eds.) Intelligent Virtual Agents, vol 10011, pp 283–294. Springer International Publishing, Cham (2016). https://doi.org/10.1007/978-3-319-47665-0_25
29. Greenwald, A., Jennings, N.R., Stone, P.: Agents and markets. IEEE Intell. Syst. **18**(6), 12–14 (2003). https://doi.org/10.1109/MIS.2003.1249164
30. Haberland, V., Miles, S., Luck, M.: Adaptive negotiation for resource intensive tasks in grids. In: Frontiers in Artificial Intelligence and Applications, 241st edn., pp. 125–136 (2012)
31. Hernandez-Leal, P., Kaisers, M., Baarslag, T., Cote, E.M. de: A Survey of Learning in Multiagent Environments. Dealing with Non-Stationarity (2017)
32. Hindriks, K.V., Jonker, C.M., Tykhonov, D.: A Multi-agent environment for negotiation. In: El-Fallah-Seghrouchni, A., Dix, J., Dastani, M., Bordini, R.H. (eds.) Multi-Agent Programming, pp. 333–363. Springer, Boston, MA (2009). https://doi.org/10.1007/978-0-387-89299-3_10
33. Jia-hai, Y., Shun-kun, Y., Zhao-guang, H.: A multi-agent trading platform for electricity contract market. In: International Power Engineering Conference, 1024–1029, vol. 2. IEEE (2005). https://doi.org/10.1109/IPEC.2005.207058
34. Kamath, S., D'Souza, R.: An ANN based classification scheme for QoS negotiation to achieve overall social benefit in cloud computing. Int. J. Appl. Eng. Res. 2262–2271 (2017)
35. Kelman, H.C.: Negotiation as interactive problem solving. Int. Negot. **1**(1), 99–123 (1996). https://doi.org/10.1163/157180696X00313
36. Kersten, G.E., Lai, H.: Negotiation support and e-negotiation systems: an overview. Group Decis. Negot. **16**(6), 553–586 (2007). https://doi.org/10.1007/s10726-007-9095-5
37. Kersten, G.E., Vahidov, R., Gimon, D.: Concession-making in multi-attribute auctions and multi-bilateral negotiations: theory and experiments. Electron. Commer. Res. Appl. **12**(3), 166–180 (2013). https://doi.org/10.1016/j.elerap.2013.02.002
38. Kiruthika, U., Somasundaram, T.S., Raja, S.K.S.: Lifecycle model of a negotiation agent: a survey of automated negotiation techniques. Group Decis. Negot. **29**(6), 1239–1262 (2020). https://doi.org/10.1007/s10726-020-09704-z
39. Köszegi, S., Kersten, G.: On-line/of-line: joint negotiation teaching in Montreal and Vienna. Group Decis. Negot. **12**(4), 337–345 (2003). https://doi.org/10.1023/A:1024879603397
40. Kowalczyk, R., van Bui, A.: FeNAs: a fuzzy e-negotiation agents system. In: IEEE/IAFE/INFORMS Conference on Computational Intelligence for Financial Engineering, pp. 26–29. IEEE (2000). https://doi.org/10.1109/CIFER.2000.844592
41. Lai, K.R., Lin, M.-W.: Modeling agent negotiation via fuzzy constraints in e-business. Comput. Intell. **20**(4), 624–642 (2004). https://doi.org/10.1111/j.0824-7935.2004.00257.x
42. Lai, K.R., Lin, M.W., Yu, T.J.: Learning opponent's beliefs via fuzzy constraint-directed approach to make effective agent negotiation. Appl. Intell. **33**(2), 232–246 (2010). https://doi.org/10.1007/s10489-009-0162-2

43. Lewicki, R.J., Barry, B., Saunders, D.M.: Negotiation, 6th edn. McGraw-Hill/Irwin, Boston (2010)
44. Lewicki, R.J., Saunders, D.M., Barry, B.: Negotiation. Readings, Exercises and Cases, 6th edn. McGraw-Hill, New York (2010)
45. Lin, R., Oshrat, Y., Kraus, S.: Investigating the benefits of automated negotiations in enhancing people's negotiation skills. In: 8th International Conference on Autonomous Agents and Multiagent Systems, pp. 345–352. IFAAMAS, Richland, SC (2009)
46. Loewenstein, J., Thompson, L.: Learning to negotiate. Novice and experienced negotiators. In: Thompson, L. (ed.) Negotiation Theory and Research, pp. 77–97 (2006)
47. Lopes, F., Novais, A.Q., Mamede, N., Coelho, H.: Negotiation among autonomous agents. Experimental evaluation of integrative strategies. In: Purtuguese Conference on Artificial Intelligence, pp. 280–288. IEEE (2005). https://doi.org/10.1109/EPIA.2005.341231
48. Ludwig, S.A.: Agent-based assistant for e-negotiations. In: An, A., Matwin, S., Raś, Z.W., Ślęzak, D. (eds.) Foundations of Intelligent Systems. LNCS (LNAI), vol. 4994, pp. 514–524. Springer, Heidelberg (2008). https://doi.org/10.1007/978-3-540-68123-6_56
49. Luo, X., Jennings, N.R., Shadbolt, N., Leung, H., Lee, J.H.: A fuzzy constraint based model for bilateral, multi-issue negotiations in semi-competitive environments. Artif. Intell. **148**(1–2), 53–102 (2003). https://doi.org/10.1016/S0004-3702(03)00041-9
50. Masvoula, M., Kanellis, P., Martakos, D.: A review of learning methods enhanced in strategies of negotiating agents. In: 2nd ICEIS, pp. 212–219 (2010)
51. Melzer, P., Schoop, M.: The effects of personalised negotiation training on learning and performance in electronic negotiations. Group Decis. Negot. **25**(6), 1189–1210 (2016). https://doi.org/10.1007/s10726-016-9481-y
52. Meyer, M.: an explorative study of the usage of negotiation styles in higher education. In: UK Academy for Information Systems Conference (2022)
53. Miller, O.: The negotiation style: a comparative study between the stated and in- practice negotiation style. Procedia. Soc. Behav. Sci. **124**, 200–209 (2014). https://doi.org/10.1016/j.sbspro.2014.02.478
54. Mirzayi, S., Taghiyareh, F., Nassiri-Mofakham, F.: An opponent-adaptive strategy to increase utility and fairness in agents' negotiation. Appl. Intell. **52**(4), 3587–3603 (2022). https://doi.org/10.1007/s10489-021-02638-2
55. Nickerson, R.C., Varshney, U., Muntermann, J.: A Method for taxonomy development and its application in information systems. Eur. J. Inf. Syst. **22**(3), 336–359 (2013). https://doi.org/10.1057/ejis.2012.26
56. Paliwal, A., Nabil, A., Atluri, V. et al.: Electronic negotiation of government contracts through transducers. In: Annual National Conference on Digital Government Research, pp. 1–6 (2003)
57. Park, J., Rahman, H.A., Suh, J. et al.: A study of integrative bargaining model with argumentation-based negotiation. Sustainability **11**(23), 6832 (2019). https://doi.org/10.3390/su11236832
58. Patel, D., Gupta, A.: An adaptive negotiation strategy for electronic transactions. In: 10th International Enterprise Distributed Object Computing Conference, pp. 243–252. IEEE (2006). https://doi.org/10.1109/EDOC.2006.14
59. Patrikar, M., Vij, S., Mukhopadhyay, D.: An approach on multilateral automated negotiation. Proc. Comput. Sci. **49**, 298–305 (2015). https://doi.org/10.1016/j.procs.2015.04.256
60. Rahwan, I., McBurney, P., Sonenberg, L.: Towards a theory of negotiation strategy (a preliminary report). In: 5th Workshop on Game Theoretic and Decision Theoretic Agents (2003)
61. Rahwan, I., Sonenberg, L., Jennings, N.R., McBurney, P.: Stratum: a methodology for designing heuristic agent negotiation strategies. Appl. Artif. Intell. **21**(6), 489–527 (2007). https://doi.org/10.1080/08839510701408971

62. Raiffa, H., Richardson, J., Metcalfe, D.: Negotiation Analysis. The Science and Art of Collaborative Decision Making. Belknap Press of Harvard University Press, Cambridge (2002)
63. Ross, L., Stillinger, C.: Barriers to conflict resolution. Negot. J. **7**(4), 389–404 (1991). https://doi.org/10.1111/j.1571-9979.1991.tb00634.x
64. Schoop, M.: Support of complex electronic negotiations. In: Marc Kilgour, D., Eden, C. (eds.) Handbook of Group Decision and Negotiation, pp. 409–423. Springer Netherlands, Dordrecht (2010). https://doi.org/10.1007/978-90-481-9097-3_24
65. Schoop, M., Jertila, A., List, T.: Negoisst: a negotiation support system for electronic business-to-business negotiations in e-commerce. Data Knowl. Eng. **47**(3), 371–401 (2003). https://doi.org/10.1016/S0169-023X(03)00065-X
66. Shi, B., Sim, K.M.: A Regression-based coordination for concurrent negotiation. In: International Symposium on Electronic Commerce and Security, pp. 699–703. IEEE (2008). https://doi.org/10.1109/ISECS.2008.114
67. Sim, K.M.: A survey of bargaining models for grid resource allocation. SIGecom Exch. **5**(5), 22–32 (2006). https://doi.org/10.1145/1124566.1124570
68. Sim, K.M.: Unconventional Negotiation: Survey and New Directions. In: ICEB (2009)
69. Sim, K.M.: Grid resource negotiation: survey and new directions. IEEE Trans. Syst. Man Cybern. Part C (App. Rev.) **40**(3), 245–257 (2010). https://doi.org/10.1109/TSMCC.2009.2037134
70. Sim, K.M., Wang, S.Y.: Designing flexible negotiation agent with relaxed decision rules. In: IEEE/WIC International Conference on Intelligent Agent Technology, pp. 282–289. IEEE Computer Society (2003). https://doi.org/10.1109/IAT.2003.1241080
71. Thomas, K.W., Kilmann, R.H.: Thomas-kilmann conflict mode instrument. Group. Organ. Stud. **1**(2), 249–251 (1976)
72. Thomas, K.W., Kilmann, R.H.: Thomas-Kilmann Conflict Mode. Profile and Interpretitive Report (2008)
73. Thompson, L.L.: The Mind and Heart of the Negotiator. Pearson, Harlow, England (2022)
74. Vahidov, R., Kersten, G., Saade, R.: An experimental study of software agent negotiations with humans. Decis. Support Syst. **66**(3), 135–145 (2014). https://doi.org/10.1016/j.dss.2014.06.009
75. Vahidov, R., Saade, R., Yu, B.: Effects of negotiation tactics and task complexity in software agent. In: Ishida, T., et al. (eds.) 18th Annual International Conference on Electronic Commerce e-Commerce in Smart connected World, pp. 1–5. ACM Press, New York, New York, USA (2016). https://doi.org/10.1145/2971603.2971620
76. Vahidov, R., Kersten, G., Yu, B.: Human-agent negotiations. The impact agents' concession schedule and task complexity on agreements. In: Hawaii International Conference on System Sciences (2017)
77. Vahidov, R., Saade, R., Yu, B.: The effects of interplay between negotiation tactics and task complexity in software agent to human negotiations. Electron. Commer. Res. Appl. **26**(3), 50–61 (2017). https://doi.org/10.1016/j.elerap.2017.09.007
78. Vetschera, R., Kersten, G., Koeszegi, S.: User Assessment of internet-based negotiation support systems: an exploratory study. J. Organ. Comput. Electron. Commer. **16**(2), 123–148 (2006). https://doi.org/10.1207/s15327744joce1602_3
79. Vetschera, R., Filzmoser, M., Mitterhofer, R.: An analytical approach to offer generation in concession-based negotiation processes. Group Decis. Negot. **23**(1), 71–99 (2012). https://doi.org/10.1007/s10726-012-9329-z
80. Vij, S.R., Mukhopadhyay, D., Agrawal, A.J.: Automated negotiation in e commerce: protocol relevance and improvement techniques. In: 61th Computers Materials & Continua, pp. 1009–1024 (2019)

81. Vom Brocke, J., Simons, A., Niehaves, B., Reimer, K., Plattfaut, R., Cleven, A.: Reconstructing the giant. On the importance of rigour in documenting the literature search process. In: European Conference on Information Systems (2009)
82. Vom Brocke, J., Simons, A., Riemer, K., Niehaves, B., Plattfaut, R., Cleven, A.: Standing on the shoulders of giants. In: Challenges and Recommendations of Literature Search in Information Systems Research. CAIS, vol. **37** (2015). https://doi.org/10.17705/1CAIS.03709
83. Wachowicz, T., Wu, S.: Negotiators' strategies and their concessions. In: Conference on Group Decision and Negotiation, pp. 254–259 (2010)
84. Walton, R.E., MacKersie, R.B.: A Behavioral Theory of Labor Negotiations. An Analysis of a Social Interaction System, 2nd edn. McGraw-Hill, New York (1991)
85. Westbrook, K.W.: Risk coordinative Maneuvers during buyer-seller negotiations. Ind. Mark. Manage. **25**(4), 283–292 (1996). https://doi.org/10.1016/0019-8501(95)00130-1
86. Yinping, Y., Singhal, S.: Designing an intelligent agent that negotiates tact-fully with human counterparts: a conceptual analysis and modeling frame-work. In: 42th Hawaii International Conference on System Sciences, pp 1–10. IEEE (2009). https://doi.org/10.1109/HICSS.2009.149
87. Zulkernine, F.H.: A Comprehensive Service Management Middleware for Autonomic Management of Composite Web Services-based Processes. Canadian theses. Library and Archives Canada, Ottawa (2011)
88. Zulkernine, F.H., Martin, P.: An adaptive and intelligent SLA negotiation system for web services. IEEE Trans. Serv. Comput. **4**(1), 31–43 (2011). https://doi.org/10.1109/TSC.2010.44

The More the Merrier? A Machine Learning Analysis of Information Growth in Negotiation Processes

Muhammed-Fatih Kaya(✉) and Mareike Schoop

University of Hohenheim, 70599 Stuttgart, Germany
{Muhammed-Fatih.Kaya,Schoop}@uni-hohenheim.de

Abstract. The exchange of information is an essential means for being able to conduct negotiations and to derive situational decisions. In electronic negotiations, information is transferred via the communication channel in the form of requests, offers, questions, and clarifications. Taken together, such information makes or breaks the negotiation. Whilst information analysis has traditionally been conducted through human coding, machine learning techniques now enable automated analyses. One of the grand challenges of e-negotiation research is the generation of future-oriented predictions whether ongoing negotiations will be accepted or rejected at the end of the negotiation process by considering the previous negotiation course. With this goal in mind, the present research paper investigates how predictive machine learning models react to the successive increase of negotiation data. Information in different data combinations is used for the evaluation of classification techniques to simulate the progress in negotiation processes and to investigate the impact of utility and communication data. It will be shown that the more information the merrier does not always hold. Instead, data-driven ML model recommendations are presented as to when and based on which data density certain models should or should not use for the analysis of electronic negotiations.

Keywords: Machine Learning · Information Growth · Negotiation Outcome · Classification · Prediction Performance · Model Selection

1 Motivation

The use of information is of particular importance in business interactions, regardless of whether it is exchanged via digital or analogue channels. Especially in the data age, where a gigantic stream of data exists, humans need information to understand facts, reflect on them and derive strategic decisions by processing valuable information [11, 25]. The exchange of information also plays a central role in the application area of electronic negotiations. Negotiators adjust to the negotiation partner and the process by considering the information situation, prioritise their issues, and plan their strategies or rather tactics to be able to use them as negotiation actions [29, 31]. Through active interaction in the form of offers, counteroffers, arguments, further information is exchanged. These exchanges take place until the defined goal of the negotiation is achieved and the

Y. Maemura et al. (Eds.): GDN 2023, LNBIP 478, pp. 20–34, 2023.
https://doi.org/10.1007/978-3-031-33780-2_2

negotiation is either accepted or rejected [43, 56]. Even though negotiations are very dynamic and can have different trajectories, negotiators have always been interested in whether the positive or negative end of negotiations can be predicted by considering the available information [53, 58]. Such reliable prediction would be an important strategic metric for negotiators and could provide important strategic indicators.

The use of machine learning (ML) can assist in achieving this goal. However, in addition to the application of intelligent algorithms, ML methods require sufficient historical data containing rich information [15]. Negotiation processes generate additional data as the negotiation progresses. This increase in data leads to the question to what extent ML methods can generate reliable predictions regarding the acceptance or rejection of negotiations and whether there is a difference between the underlying negotiation data (utility data vs. communication data). The dimension of negotiation progress should not be neglected either, i.e. the negotiation section at which sufficient information is available to be able to derive valid predictions. The investigation of the influence of gradually increasing information on the prediction performance of ML models is unexplored in the electronic negotiation context considering communication data and utility data. The present research paper aims to address this problem by investigating how selected ML methods respond to the increase in negotiation information and how well the outcome of negotiations can be predicted considering the data combinations. Our research question is thus:

What impact does the successive increase in negotiation information have on the performance of ML models for predicting accepted and rejected negotiations?

The particular importance of information in electronic negotiations and the application potential of ML are described and the methodological background is introduced in Sect. 2. The evaluation of the predictive power of renowned ML methods is performed in Sect. 3 by simulating the increase of negotiation information. Section 4 derives implications and discusses a critical reflection of the value contribution of the results. Finally, core statements of the paper are summarised and a research outlook is presented in the final chapter.

2 Theoretical Background

2.1 The Exchange of Information in Negotiations

"Negotiations are iterative communication and decision-making processes between two or more parties who cannot achieve their objectives through unilateral actions, exchange information comprising offers, counteroffers and arguments, deal with interdependent tasks and search for a consensus" [8]. These interactions take place until the negotiating parties either agree for a given offer or reject it due to a lack of consensus [43]. Nowadays, electronic negotiation systems are used to conduct negotiations, which ensure the exchange of information via electronic media [8]. Negotiation Support Systems (NSSs) represent one form of electronic negotiation systems. They pursue the goal of supporting the negotiating parties in decision-making, communication management and document management of negotiations [51].

The NSS Negoisst, whose data is used in this paper, serves this goal. It follows a defined negotiation protocol, specifies negotiation issues and ensures the structured exchange of negotiation messages [50, 51]. In the negotiation process, parties exchange offers and counteroffers until the last offer is either rejected or accepted. Negotiators submit priorities of negotiation issues that are expressed in utility metrics and have the possibility to evaluate various issue-based combinations by using the multi attributive utility theory (MAUT) during this process [47]. The Negoisst system generates four utility metrics: (1) individual utility of the offer sender, (2) individual utility of the offer receiver (3) joint utility that represents the total utility of sender and receiver, and (4) the contract imbalance which presents a control measure for fairness [48, 50].

In addition to these metrics, textual communication messages are exchanged through the use of natural language [49]. Negotiation communication involves the exchange of information, conveys insight into the intentions of the negotiating partner and enables the use of strategies [2, 44]. This allows additional information to be shared which would be missing through the sole exchange of utilities. In summary, each of the exchanged negotiation offers contains two central channels of information: utility data and communication data.

The exchange of information is essential for negotiating parties to move the negotiation forward and to be able to derive well-founded strategic decisions [66]. Information disclosure is often perceived as a sign of trust for the other party and is the basis for any win-win situation [57]; information disclosure leads to higher levels of joint outcome [61]. However, the mere sharing of information is not sufficient; the content of the information is crucial. For example, the disclosure of preferences is perceived as a positive sign by the negotiation partner; information about a lack of negotiation possibilities might have negative effects on the negotiation climate [22, 57]. Therefore, information content must be considered. Integrative information and distributive information can be distinguished [39, 60]. To give an example, negotiations focusing on distributive outcomes mostly contain the exchange of positional information; negotiations being characterised by integrative outcomes contain information on priorities [41]. Notwithstanding the various information orientations, sharing information contributes to a better mutual understanding in most cases [45, 69].

Information is exchanged with varying intensity starting from the first interactions up to the negotiation end by using natural language and specifying utility-specific preferences [50]. The exchange of information regarding e.g. interests and preferences is of particular importance at the beginning of a negotiation, as this can reduce the asymmetric distribution of information and achieve a better negotiation outcome [73]. Negotiation researchers refer to this as the problem identification phase where a high-level exchange of e.g., preference information takes place [1, 14]. Options for solving the conflict problem, interests as well as intentions are discussed, and attempts are being made to create mutual trust [40]. However, the exchange of information decreases during negotiation process and the closer negotiators get to the final deadline [1, 32, 37]. The use of negotiation information is an important element if the entire process is considered as it enables the negotiators to identify a clear tendency for the further course of the negotiation and to assess the extent to which the further process will be promising or not [33].

2.2 Methodological Background

The use of ML methods in electronic negotiations pursues the overarching goal of identifying systematic patterns in large amounts of data in order to derive value-adding predictions for the negotiation process [70]. Predictive insights can provide significant strategic value by aligning subsequent negotiator decisions with trends in predictive metrics. The effort to examine electronic negotiation data systematically through ML methods has existed for years. So far, ML methods have been used in the context of negotiations, e.g., to classify the use of strategies and tactics, to study the use of emotions in negotiation language or to predict offer-based preferences of negotiation counterparties [9, 16, 53]. Predictive analyses were conducted based on either utility or communication data. A combined analysis of the data with regard to the question of how the successive increase in negotiation information influences the performance of predictive ML, has remained unexplored so far. This will be investigated in this paper by using classification approaches.

Classification approaches are assigned to supervised learning methods. They train based on the underlying data to predict a defined target set of classes (in our case: accepted or rejected negotiations) [12]. In this paper, four renowned classification approaches are used for this purpose. The methods are evaluated using different combinations of utility and communication data and are subsequently simulated by the gradual increase of information sections (see Sect. 3.1). For the iterative improvement of the models, hyperparameter optimisations are performed to determine the best parameter combinations for the maximisation of prediction accuracies. Here, the n-fold cross validation approach is used for all ML-methods. Cross validation helps to obtain a reasonable set of training data points for the development of the models and ensures a balanced evaluation [10, 38]. Training data is divided into n-equally sized partial data sets, while models are trained using the n-1 partial data sets. The n-th data set validates the trained model by calculating the accuracy of the particular partition. This step is repeated until each subset has served as a validator in one of the iterations and the average of all accuracies is calculated as performance measure after the last iteration [64].

The Support Vector Machine (SVM) [19, 21] represents the first method which is going to be evaluated for the successive growth of negotiation information. It is a commonly used method for the classification of binary states and is particularly suitable for the classification of textual data [17]. The goal of SVM is to position a so-called hyperplane in the vector space of data so that the underlying data can be divided into two classes as precisely as possible [34]. To find the perfect hyper-plane to separate the classes, the algorithm maximises the span between the hyperplane and the closest point of both classes in the vector space [24]. The major advantage of an SVM is that it can be applied to complex data sets. Hence, it can provide a very good answer to the curse of dimensionality issues in machine learning [19, 21].

The second approach used is Naive Bayes (NB) [65], which is based on the Bayes theorem and uses an underlying probability model. This model provides a probability for each instance belonging to one of the defined classes [23, 36]. The overall probability distribution of all attributes is used for the final classification, since the attributes influence each other and an optimal decision can only be derived by iteratively optimising the relative frequencies of all instances [68]. NB is among the simplest algorithms and yields good results on numerical as well as textual datasets [63, 65]. Moreover, it proves to be highly efficient in terms of training time and of classification running time [5].

The k-nearest neighbour method (kNN) [71] is a non-parametric model that performs a simple function on the training data set. The model searches for the k-nearest neighbours in the training. For the classification of a new instance, the closest instance that has similar characteristic properties is always used [54, 71]. A unique feature of kNN is that it does not include a classical training process which we have with other methods. Hence, the training data is stored in the training phase only for the purpose of later predictions of new datasets, so that the learning effort is omitted [74]. The search for the most similar dataset at the time of classification requires all stored data to be checked in terms of their distance from the new dataset to be classified [46].

Finally, a decision tree (DT) [42] is used as the last predictive ML method to be evaluated. DTs classify instances by sorting them based on feature values. Each node in a DT represents a feature of an object to be classified, and each branch represents a value that the node can assume. Instances are classified starting at the root node and sorted based on their feature values [30]. The feature that divides the training data best is defined as the root node of the tree. Hence, the chains of a tree path starting from the root node to one of the leaves can be expressed in the form of if-then rules. One of the most useful characteristic of decision trees is their comprehensibility [3]. People can easily understand why a decision tree classifies an instance as belonging to a specific class. Moreover, DTs represent an efficient method in terms of their training and processing time [72]. They are able to process both categorical and numerical input in a natural way which simplifies the preparation phase of data [42, 52].

3 Results

3.1 Data Processing

The experimental part of this research paper uses a balanced negotiation dataset (i.e. equally distributed number of messages and accordingly offers between accepted and rejected negotiations) of the Negoisst system with 2232 messages from 215 completed negotiations. This dataset was collected as part of ten student negotiation experiments between 2010 and 2016. Those experiments were selected which did not contain manipulations. Each message contains metric utility data in addition to textual communication data. Both data types have to be processed before they can be integrated for the training of the presented classification methods. Textual communication data in particular requires special treatment and has to be prepared for the generation of a processable vector space model by using elaborate Text Mining methods [55].

The processing includes the following steps (1) tokenising, (2) filtering and stop-word removal, (3) n-gram generation (4) stemming and (5) dimensionality reduction. The messages are (1) divided into word units; (2) frequently occurring grammatical words like "the", "is", "are" are removed; (3) regularly co-existing word combinations are generated; (4) the words are returned to their root words using the Porter stemmer to avoid word constellations [4, 59]. These elaborate processing milestones and the application of the TF-IDF weighting approach [67] lead to a communication vector with 5929 word dimensions: [2232 negotiation messages x 5929 dimensions]. To counteract the phenomenon of high-dimensionality and to achieve a more compact word representation, a dimension reduction method called Optimise Selection is applied in the last processing step [26]. This leads to the sole consideration of those dimensions of negotiation messages (2976 dimensions) which significantly contribute to the prediction accuracy of accepted and rejected negotiations.

Three datasets are formed in the next step to measure the successive increase of information based on the underlying data types: (a) dataset with exclusively four utility data information presented in Sect. 2.1: [2232 negotiation messages x 4 dimensions], (b) dataset with exclusively communication data: [2232 negotiation messages x 2976 dimensions], and (c) an entire dataset containing (a) and (b): [2232 negotiation messages x 2980 dimensions]. Therefore, the influence of different negotiation information can be investigated in performance evaluations. Subsequently, negotiation messages are subjected to descriptive analysis and are discretised into four equally sised section classes by the binning technique to evaluate the influence of successive information increase on ML models by considering the dimensions of negotiation progress [35]. Four sections are generated which contain those messages that were exchanged up to the given negotiation progress: 25% of the negotiation progress (section I), 50% of the negotiation progress (section II), 75% of the negotiation progress (section III), and completed negotiations (section IV). It is important to note that the individual progress sections do not represent a time dimension. Those messages which lie at the section boundary (due to the number of message not being divisible by four) are assigned to the closest section. Four section-based data sets are generated for each data combination (a) to (c) depending on the progress of negotiations (I) to (IV). Thus, a total of 12 complementary data sets could be generated which serve as data input for ML.

3.2 Evaluation Results of Model-Based Prediction Performance

As introduced in Sect. 2.2, four classification methods are used to evaluate the performance of ML models regarding the prediction of accepted and rejected negotiations. This diverse use of methods serves the goal of minimising the algorithmic bias and maximising the predictive power of ML models. To maximise the generalisability of predictions, cross-validation is used for the execution of efficient training and testing as described in Sect. 2.2. Each of the data combinations (a) to (c), and sections (I) to (IV) (starting from a 25% information share to the completed negotiation) are gradually trained to measure the impact of increasing negotiation information. The prediction performance is measured via the F1-score which represents a renowned metric for classification-based predictions [18].

Initially, the classification methods are evaluated by considering the sections and the prediction objective for the exclusive usage of utility data (see Fig. 1). As can be seen in the line chart, all four methods show different conspicuities. The trained SVM model shows the highest performance value with an F1-score of 0.63 for the information exchanged in the first 25% of the negotiation. A linear increase of the performance can be observed in the further course with F1-scores of 0.66 (50% of negotiation) and 0.67 (75% of negotiation). Finally, this gradual F1-score increase leads to a score of 0.69 for completed negotiations. The NB model starts with the worst result in the first section with an F1-score of 0.38 for 25% of the negotiation and significantly increases to an

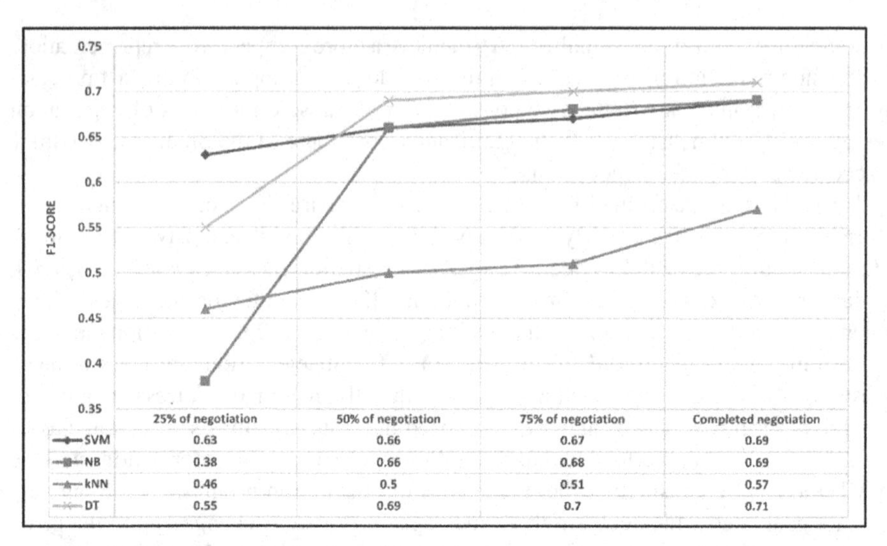

Fig. 1. Performance Evaluation of Classification Models based on (a) Utility Data

F1-score of 0.66 in the second section until a performance of 0.69 can be achieved for completed negotiations. Similar to the NB model, the DT model starts with an F1-score of 0.55, increases to 0.69 in the second section and outperforms all other models with an F1-score of 0.71. Even though an improvement can be achieved with increasing information on each section, the kNN model does not manage to outperform the other classification methods (see Fig. 1). Starting with an F1-score of 0.46 in section 1, this model scores 0.5 and 0.51 in the second and third sections and ends with an F1-score of 0.57 for completed negotiations. These results show that the performance of the kNN model differs from the others in a negative way. While the SVM model achieves good performance values from the beginning of negotiation with 25% of negotiation data, the models NB, kNN and DT generate the largest increase in performance after processing 50% of the utility data. E.g., a rapid increase to an F1-score of 0.66 can be observed for the NB model in the second section of the negotiation. The DT model outperforms all other models until the completion of the negotiation after 50% of utility information is provided (see Fig. 1).

In Fig. 2, only processed communication data are considered for the section-by-section evaluation. While the F1-score of the DT model remains constant at 0.6 in the first two sections, an improvement in performance can only be observed with 75% of the communication information. A score of 0.65 can be finally achieved for completed negotiations. The models for SVM, NB and kNN initially show similar results for section 1 in the interval from 0.64 (for the kNN model) to 0.66 (for the NB model). The performance for the SVM model increases gradually with increasing communication information by starting with an F1-score of 0.65 in section 1, gradually rises to 0.67 (for 50% of the negotiations) and 0.69 (for 75% of the negotiations) until a score of 0.72 is achieved for completed negotiations. The F1-score for the NB model increases from 0.66 to 0.67 in the transition from section 1 to section 2, decreases after the second section to an F1-score of 0.66 and stagnates at this point (see Fig. 2). The kNN model starts with an

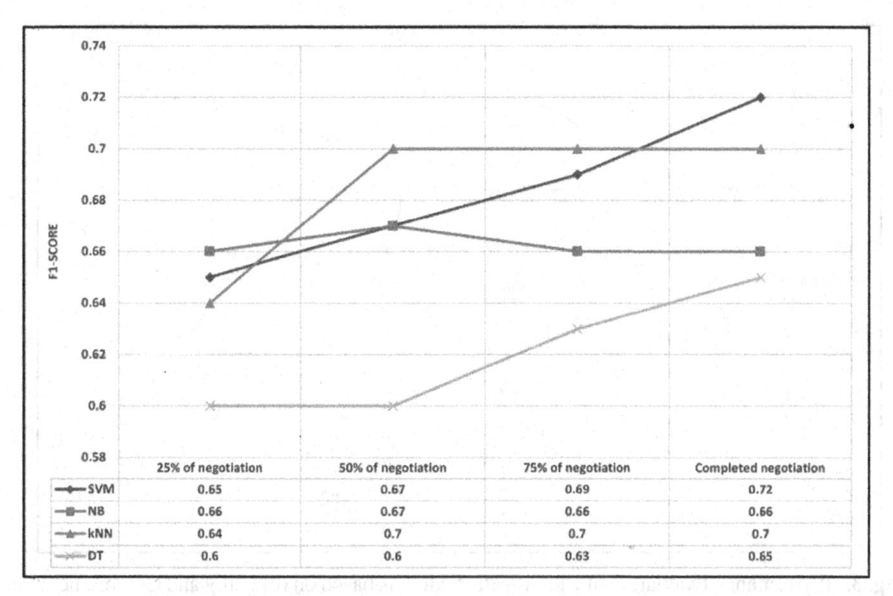

	25% of negotiation	50% of negotiation	75% of negotiation	Completed negotiation
SVM	0.65	0.67	0.69	0.72
NB	0.66	0.67	0.66	0.66
kNN	0.64	0.7	0.7	0.7
DT	0.6	0.6	0.63	0.65

Fig. 2. Performance Evaluation of Classification Models based on (b) Communication Data

F1-score of 0.64 and shows the largest performance improvement with a 25% increase in communication information between sections 1 and 2 to 0.70. This performance level can be maintained for the second half of the negotiations until the end of negotiation. The results of the communication-based evaluation show that the DT model performs worst across all sections and has difficulties with predicting the negotiation outcome by solely considering negotiation communicant data. Even though the results of the NB model are not as weak as those of the DT model, a decline in performance with and stagnation can be observed after 50% of the negotiation. The kNN model, on the other hand, shows the best predictive performance for 25% and 75% of the negotiation until it is outperformed by the gradually improving SVM model in section 4.

Utility data are examined together with the communication data for each of the sections in the third step. The results show that the performance values for the entire negotiation are between 0.65 and 0.66 for the two models NB and kNN. The DT model deviates from this observation and starts with an F1-score of 0.57 for the first section of negotiations. By adding more information (from half of the negotiation), the DT model improves the performance to an F1-score of 0.66 for the second section and performs comparable to other models in the further course of the negotiation until a score of 0.66 is reached for completed negotiations (see Fig. 3). The results of the SVM model differ with a much more positive trend. The results show that the SVM dominates all other models by gradually increasing its performance starting with an F1-score of 0.66 for 25%. The performance increases steadily with increasing negotiation information until it reaches its maximum for completed negotiations with an F1-score of 0.72. The results show that the DT model needs at least 50% of negotiation information to derive comparable results as the other models. The NB and kNN models, on the other hand, stagnate in the interval of 0.65 to 0.66 for the entire negotiation, even if further section-based information

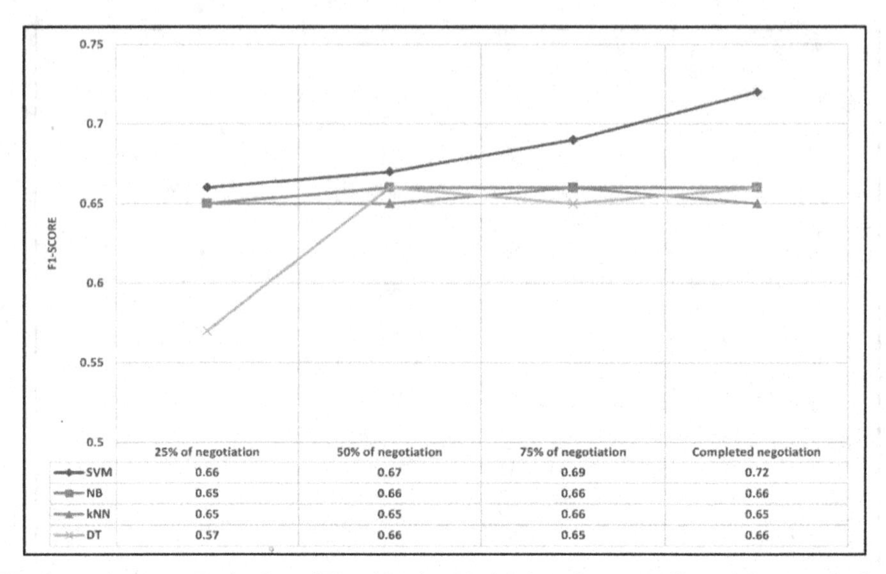

Fig. 3. Performance Evaluation of Classification Models based on (c) Utiliy and Communication Data

is provided. The SVM model represents the only model which outperforms all other models (with already 25% of the information) and gradually improves its performance with increasing negotiation information.

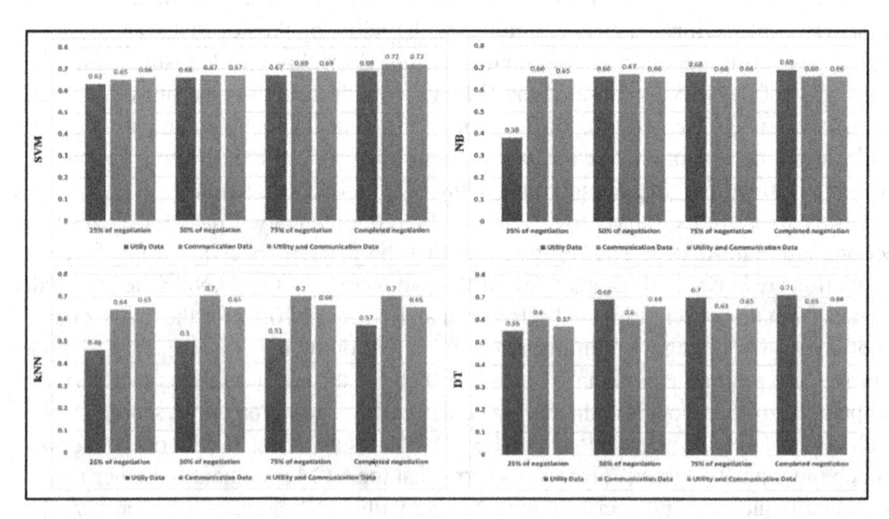

Fig. 4. Summary of Section-Based Performance Evaluation of all Data Combinations

Finally, a model-based view taking the different data combinations into account is needed to derive complementary findings regarding the prediction performance of ML (see Fig. 4). The SVM model shows that the performances do not differ significantly in the internal comparison between data combinations (a) utility data, (b) communication data and (c) as combination of (a) and (b) across all sections. A small improvement of the performance can be achieved only when considering only (b) or (c). The SVM model thus provides robust predictions for each of the sections. The NB model, on the other hand, shows a breakdown in performance in the first section for (a). An F1-score of 0.65 to 0.69 can be maintained for all sections for (b) and (c). Hence, considering (c) does not add any value to the predictive power. On the contrary, better results could be derived for the combinations (a) and (b) in sections 2 to 4. The results of kNN show that it performs worst in all sections for data set (a). In contrast, the pure consideration of (b) from section 2 up to the end achieves the best performance with an F1-score of 0.7. This result clearly distinguishes the results of (c). The DT model shows that considering only (a) from section 2 outperforms both other data combinations. Data sets (b) and (c) provide no additional value for predicting the end of the negotiation compared to data combination (a).

4 Discussion

The results of this paper clearly show that the answer to the overarching research question, what impact the successive increase in negotiation information have on the performance of ML for predicting accepted and rejected negotiations, has to be investigated from different perspectives. In addition to the temporal dimension of the information increase, the value-adding contribution of utility and communication data was considered. The question which information extension (section and data type) performance improvements can be achieved is as important as the question at which point ML models start to stagnate. Depending on the perspective, it can be deduced for ML that the models either need more information to derive a solid predictive power or that the addition of more information does not generate any additional value or even represents a disruptive factor. A comparative analysis of the three sectional curves (see Figs. 1, 2, and 3) reveals that different performances could be generated for each of the underlying data sets (a) to (c). While linear or leaping improvements in performance can be achieved for some of the evaluated models a with increasing section-based information, models exist whose performance stagnates or even decreases despite the addition of further information. This phenomenon can also be observed when considering further types of negotiation data (see Fig. 4). Taking this observation into account, it can be seen that the DT model is not robust compared to the other models. Solid predictions could only be derived for the utility-based dataset. The dependence of this model with respect to the underlying data as well as the conditional generalisability stands out negatively for our application context of electronic negotiations [20].

On the other hand, the results of the kNN clearly show that better results can be derived for the prediction of accepted or rejected negotiations when only negotiation communication is integrated. The influence of the data difference is less noticeable in the NB model. Rather, the section-based increase in communication and utility data influences the performance of the model for the second half of the negotiation. The results show that the performance of this model stagnates and even decreases after a negotiation progress of 50% (except for (a)), despite the provision of additional information (see

Figs. 2 and 3). These results implicate the fact that the increase of information does not always lead to an improvement of the predictive power in the negotiation context. General statements of this kind should be avoided and a differentiated, data-driven perspective of ML methods should be taken. Contrary to these results, the SVM model can generate better predictions with increasing information, both from the perspective of sections (I) to (IV) and from the perspective of data types (a) to (c). A linear increase of prediction performance can be observed for all of these cases (see Figs. 1, 2, and 3). Furthermore, SVM is already able to achieve better prediction results starting from the first section with 25% of available information in a cross-method comparison. ML-research showed that SVM models require less training data in terms of data adequacy to achieve the same predictive performance as comparable algorithms [6]. Better SVM-results could be derived as early as possible with regard to our evaluation results. Nevertheless, the best results for the SVM could be generated for the combined data set with utility and communication data, even if there is no great difference in performance between the sole consideration of communication data (see Fig. 4). This shows that SVM models can already derive comparatively good results on the basis of communication data [62].

Considering all results of our study, it should be noted that the largest performance improvements occur with an increase in information from section 1 to section 2. Most of the models require at least half of the negotiation information to be able to predict the future-oriented trend of the negotiation in terms of accepted or rejected negotiations. Since there is a high degree of exchange of decision-relevant information during the problem identification phase and positioning phase, which represent in the initial phase of negotiations, this might have had an influence on the ML models [1, 40]. Only linear or no improvements can be achieved in the second half of the negotiation for the majority of models. This supports the finding that information sharing steadily decreases as the negotiation process progresses and the negotiation deadline approaches [32, 37].

5 Summary and Outlook

This paper provides key research contributions for researchers analysing the digitalisation of negotiations and the application potentials of ML methods in electronic negotiations. A nuanced view is taken in this paper on the increase and influence of negotiation information on predictive ML methods as different models react differently to the underlying data combinations. Hence, it has been disproved that the successive increase of negotiation information always leads to an improvement in performance regarding the negotiation context. This paper provides valuable insights and shows which models are particularly suitable for which data types. Moreover, a section-based estimation could be given on which data fraction the models show significant performance improvements. At least half of the negotiation data should be available for a large part of the models (except SVM) to be able to detect the negotiation tendency but half of the negotiation data can be enough to make reliable predictions. Conversely, it was also possible to show cases where the inclusion of additional information was perceived as a confounding factor for some of the models and led to a deterioration in performance. For some of the ML methods, it could be shown that the processing of e.g., 50% of the negotiation generates identical or even better results than the processing of all available negotiation information. Therefore, the more the merrier does not hold in general.

One of the limitations of this study is the assumption of fixed data sections simulating the growth of negotiation information. Those are more fine-grained in reality, can be of different lengths and might be further divided considering the dynamic character of negotiations. Furthermore, student data was used in the experimental part of this study, which relativises the validity of the results from another angle. Even though it has already been pointed out that the use of ML is differentiated, it should be noted that not all existing ML methods were evaluated, but only renowned ones. Nevertheless, the derived implications highlight various future research potentials. E.g., the findings of this study could serve as initial indicators for a negotiation early warning system which we also have in other comparable business fields [7, 27]. Warnings regarding undesirable future trends (e.g., rejected negotiation) could be provided in this context by incorporating predictions that are derived at a specific time section t. This would enable the implementation of negotiation actions (strategies and tactics) which could be tailored to dynamic courses of negotiations. Future negotiation events could be positively influenced by considering such predictive metrics in the decision-making process [9]. The selection of appropriate strategy sequences can be of particular importance for the negotiation success [13, 28]. A look into the future could facilitate the use of right strategic choices. The systematic use of ML reveals numerous potentials and can make central contributions supporting with descriptive and predictive support functionalities.

References

1. Adair, W.L., Brett, J.M.: The negotiation dance: time, culture, and behavioral sequences in negotiation. Organ. Sci. **16**(1), 33–51 (2005)
2. Adair, W., Brett, J.: Culture and negotiation processes. In: Gelfand, M.J., Brett, J.M. (eds.) The Handbook of Negotiation and Culture, pp. 158–176. Stanford Business Books, California (2004)
3. Almuallim, H., Kaneda, S., Akiba, Y.: Development and applications of decision trees. In: Expert Systems, pp. 53–77. Academic Press (2002)
4. Anandarajan, M., Hill, C., Nolan, T.: Text preprocessing. In: Anandarajan, A., Hill, C., Nolan, T. (eds.) Practical Text Analytics. AADS, vol. 2, pp. 45–59. Springer, Cham (2019). https://doi.org/10.1007/978-3-319-95663-3_4
5. Ashari, A., Paryudi, I., Tjoa, A.M.: Performance comparison between Naïve Bayes, decision tree and k-nearest neighbor in searching alternative design in an energy simulation tool. Int. J. Adv. Comput. Sci. Appl. (IJACSA) **4**(11) (2013)
6. Bassiouni, M., Ali, M., El-Dahshan, E.A.: Ham and spam e-mails classification using machine learning techniques. J. Appl. Secur. Res. **13**(3), 315–331 (2018)
7. Bertoncel, T., Erenda, I., Bach, M.P., Roblek, V., Meško, M.: A managerial early warning system at a smart factory: an intuitive decision-making perspective. Syst. Res. Behav. Sci. **35**(4), 406–416 (2018)
8. Bichler, M., Kersten, G., Strecker, S.: Towards a structured design of electronic negotiations. Group Decis. Negot. **12**(4), 311–335 (2003)
9. Carbonneau, R., Kersten, G.E., Vahidov, R.: Predicting opponent's moves in electronic negotiations using neural networks. Expert Syst. Appl. **34**(2), 1266–1273 (2008)
10. Chawla, K., et al.: CaSiNo: a corpus of campsite negotiation dialogues for automatic negotiation systems. In: Toutanova, K., et al. (eds.) Proceedings of the 2021 Conference of the North American Chapter of the Association for Computational Linguistics: Human Language Technologies, Virtual Conference, pp. 3167–3185. Association for Computational Linguistics, Stroudsburg (2021)

11. Citroen, C.L.: The role of information in strategic decision-making. Int. J. Inf. Manage. **31**(6), 493–501 (2011)
12. Cunningham, P., Cord, M., Delany, S. J.: Supervised learning. In: Machine Learning Techniques for Multimedia, pp. 21–49. Springer, Berlin, Heidelberg (2008). https://doi.org/10.1007/978-3-540-75171-7_2
13. Donohue, W.A.: Analyzing negotiation tactics: development of a negotiation interact system. Humam Commun. Res. **7**(3), 273–287 (1981)
14. Donohue, W.A.: Communicative competence in mediators. In: Kressel, K., Pruitt, D.G. (eds.) Mediation Research: The Process and Effectiveness of Third-Party Intervention, pp. 322–343. Jossey-Bass, San Francisco, CA (1989)
15. Figueroa, R.L., Zeng-Treitler, Q., Kandula, S., Ngo, L.H.: Predicting sample size required for classification performance. BMC Med. Inform. Decis. Mak. **12**(1), 1–10 (2012)
16. Filzmoser, M., Koeszegi, S.T., Pfeffer, G.: What computers can tell us about emotions – classification of affective communication in electronic negotiations by supervised machine learning. In: Bajwa, D., Koeszegi, S.T., Vetschera, R. (eds.) Group Decision and Negotiation: Theory, Empirical Evidence, and Application. LNBIP, vol. 274, pp. 113–123. Springer, Cham (2016). https://doi.org/10.1007/978-3-319-52624-9_9
17. Goodfellow, I., Bengio, Y., Courville, A.: Deep Learning, 1st edn. MIT Press, Cambridge (2016)
18. Grandini, M., Bagli, E., Visani, G.: Metrics for multi-class classification: an overview. arXiv preprint arXiv:2008.05756 (2020)
19. Gualtieri, J.A., Chettri, S.: Support vector machines for classification of hyperspectral data. In: IGARSS 2000. IEEE International Geoscience and Remote Sensing Symposium. Taking the Pulse of the Planet: The Role of Remote Sensing in Managing the Environment. Proceedings, vol. 2, pp. 813–815 (2000)
20. Guggari, S., Kadappa, V., Umadevi, V.: Theme-based partitioning approach to decision tree: an extended experimental analysis. In: Sridhar, V., Padma, M.C., Rao, K.A.R. (eds.) Emerging Research in Electronics, Computer Science and Technology. LNEE, vol. 545, pp. 117–127. Springer, Singapore (2019). https://doi.org/10.1007/978-981-13-5802-9_11
21. Gunn, S.R.: Support vector machines for classification and regression. ISIS Techn. Rep. **14**(1), 5–16 (1998)
22. Harinck, F., Ellemers, N.: Hide and seek: the effects of revealing one's personal interests in intra-and intergroup negotiations. Eur. J. Soc. Psychol. **36**(6), 791–813 (2006)
23. Huang, Y., Li, L.: Naive Bayes classification algorithm based on small sample set. In: 2011 IEEE International Conference on Cloud Computing and Intelligence Systems, pp. 34–39 (2011)
24. James, G.D., Witten, T., Hastie, R.: An Introduction to Statistical Learning with Applications in R. Springer, New York (2013)
25. Janssen, M., van der Voort, H., Wahyudi, A.: Factors influencing big data decision-making quality. J. Bus. Res. **70**(1), 338–345 (2017)
26. Kaya, M.F., Schoop, M.: Maintenance of data richness in business communication data. In: Proceedings of the 28th European Conference on Information Systems (ECIS 2020), An Online AIS Conference (2020)
27. Klopotan, I., Zoroja, J., Meško, M.: Early warning system in business, finance, and economics: bibliometric and topic analysis. Int. J. Eng. Bus. Manag. **10**, 1847979018797013 (2018)
28. Koeszegi, S.T., Pesendorfer, E.M., Vetschera, R.: Data-driven phase analysis of e-negotiations: an exemplary study of synchronous and asynchronous negotiations. Group Decis. Negot. **20**, 385–410 (2011)
29. Koeszegi, S. T., Vetschera, R.: Analysis of negotiation processes. In: Handbook of Group Decision and Negotiation, pp. 121–138. Springer, Dordrecht (2010). https://doi.org/10.1007/978-90-481-9097-3_8

30. Kotsiantis, S.B., Zaharakis, I., Pintelas, P.: Supervised machine learning: a review of classification techniques. Emerg. Artif. Intell. Appl. Comput. Eng. **160**(1), 3–24 (2007)
31. Lewicki, R.J., Saunders, D.M., Minton, J.W., Roy, J., Lewicki, N.: Essentials of negotiation, p. 304. McGraw-Hill/Irwin, Boston, MA, USA (2011)
32. Lim, S.G., Murnighan, J.K.: Phases, deadlines, and the bargaining process. Organ. Behav. Hum. Decis. Process. **58**(1), 153–171 (1994)
33. Lopes, F., Coelho, H.: Strategic and tactical behaviour in automated negotiation. Int. J. Artif. Intell. **4**(1), 35–63 (2010)
34. Marsland, S.: Machine Learning: An Algorithmic Perspective. CRC Press, Boca Raton (2015)
35. Marzuki, Z., Ahmad, F.: Data mining discretization methods and performances. Lung **3**(32), 57 (2012)
36. Mitchell, T.: Machine Learning. 1st edn. McGraw-Hill, New York (1997)
37. Moore, D.A.: Myopic prediction, self-destructive secrecy, and the unexpected benefits of revealing final deadlines in negotiation. Organ. Behav. Hum. Decis. Process. **94**(2), 125–139 (2004)
38. Nastase, V., Koeszegi, S., Szpakowicz, S.: Content analysis through the machine learning mill. Group Decis. Negot. **16**(4), 335–346 (2007)
39. Olekalns, M., Brett, J.M., Weingart, L.R.: Phases, transitions and interruptions: modeling processes in multi-party negotiations. Int. J. Confl. Manag. **14**(3/4), 191–211 (2003)
40. Olekalns, M., Smith, P.L.: Understanding optimal outcomes. Hum. Commun. Res. **26**(4), 527–557 (2000)
41. Olekalns, M., Smith, P.L., Walsh, T.: The process of negotiating: strategy and timing as predictors of outcomes. Organ. Behav. Hum. Decis. Process. **68**(1), 68–77 (1996)
42. Patel, H.H., Prajapati, P.: Study and analysis of decision tree based classification algorithms. Int. J. Comput. Sci. Eng. **6**(10), 74–78 (2018)
43. Pruitt, D. G.: Negotiation Behavior. Academic Press (2013)
44. Putnam, L.L., Jones, T.S.: The role of communication in bargaining. Hum. Commun. Res. **8**(3), 262–280 (1982)
45. Ram, A., Ross, H.: 'We got to figure it out': information-sharing and siblings' negotiations of conflicts of interests. Soc. Dev. **17**(3), 512–527 (2008)
46. SalvadorMeneses, J., RuizChavez, Z., GarciaRodriguez, J.: Compressed k NN: K-nearest neighbors with data compression. Entropy **21**(3), 234 (2019)
47. Scholz, R., Tietje, O.: Embedded Case Study Methods. SAGE Publications, Inc., Thousand Oaks (2002). https://doi.org/10.4135/9781412984027
48. Schoop, M., Jertila, A., List, T.: Negoisst: a negotiation support system for electronic business-to-business negotiations in e-commerce. Data Knowl. Eng. **47**(3), 371–401 (2003)
49. Schoop, M.: The role of communication support for electronic negotiations. In: Kamiński, B., Kersten, G.E., Szapiro, T. (eds.) Outlooks and Insights on Group Decision and Negotiation. LNBIP, vol. 218, pp. 283–287. Springer, Cham (2015). https://doi.org/10.1007/978-3-319-19515-5_22
50. Schoop, M.: Negoisst: complex digital negotiation support. In: Marc Kilgour, D., Eden, C. (eds.) Handbook of Group Decision and Negotiation, pp. 1149–1167. Springer International Publishing, Cham (2021). https://doi.org/10.1007/978-3-030-49629-6_24
51. Schoop, M.: Support of complex electronic negotiations. In: Kilgour, D., Eden, C. (eds.) Handbook of Group Decision and Negotiation, Advances in Group Decision and Negotiation, vol. 4, pp. 409–423. Springer, Dordrecht (2010)
52. Sharma, H., Kumar, S.: A survey on decision tree algorithms of classification in data mining. Int. J. Sci. Res. (IJSR) **5**(4), 2094–2097 (2016)
53. Sokolova, M., Lapalme, G.: How much do we say? Using informativeness of negotiation text records for early prediction of negotiation outcomes. Group Decis. Negot. **21**(3), 363–379 (2012)

54. Steinbach, M., Tan, P.N.: kNN: k-nearest neighbors. In: The Top Ten Algorithms in Data Mining, pp. 165–176. Chapman and Hall, CRC (2009)
55. Talib, R., Hanif, M.K., Ayesha, S., Fatima, F.: Text mining: techniques, applications and issues. Int. J. Adv. Comput. Sci. App. **7**(11) (2016)
56. Thompson, L.L.: Information exchange in negotiation. J. Exp. Soc. Psychol. **27**(2), 161–179 (1991)
57. Thompson, L.L., Wang, J., Gunia, B.C.: Negotiation. In: Levine, J.M. (ed.) Group Processes, pp. 55–84. Psychology Press (2013)
58. Twitchell, D.P., Jensen, M.L., Derrick, D.C., Burgoon, J.K., Nunamaker, J.F.: Negotiation outcome classification using language features. Group Decis. Negot. **22**(1), 135–151 (2013)
59. Vel, S. S.: Pre-Processing techniques of text mining using computational linguistics and python libraries. In: 2021 International Conference on Artificial Intelligence and Smart Systems (ICAIS), pp. 879–884 (2021)
60. Weingart, L.R., Brett, J.M., Olekalns, M., Smith, P.L.: Conflicting social motives in negotiating groups. J. Pers. Soc. Psychol. **93**(6), 994–1009 (2007)
61. Weingart, L.R., Thompson, L.L., Bazerman, M.H., Carroll, J.S.: Tactical behavior and negotiation outcomes. Int. J. Confl. Manag. **1**(1), 7–31 (1990)
62. Witten, I.H., Frank, E., Hall, M.A., Pal, C.J.: Data Mining: Practical Machine Learning Tools and Techniques, 4th edn. Morgan Kaufmann, Cambridge (2016)
63. Witten, I.H., Frank, E.: Data mining: practical machine learning tools and techniques with Java implementations. ACM SIGMOD Rec. **31**(1), 76–77 (2002)
64. Wong, T.T.: Performance evaluation of classification algorithms by k-fold and leave-one-out cross validation. Pattern Recogn. **48**(9), 2839–2846 (2015)
65. Xu, S.: Bayesian Naïve Bayes classifiers to text classification. J. Inf. Sci. **44**(1), 48–59 (2018)
66. Young, M.J., Bauman, C.W., Chen, N., Bastardi, A.: The pursuit of missing information in negotiation. Organ. Behav. Hum. Decis. Process. **117**(1), 88–95 (2012)
67. Yun-tao, Z., Ling, G., Yong-cheng, W.: An improved TF-IDF approach for text classification. J. Zhejiang Univ. Sci. A **6**(1), 49–55 (2005). https://doi.org/10.1007/BF02842477
68. Zaki, M.J., Meira, W.: Data Mining and Analysis: Fundamental Concepts and Algorithms. Cambridge University Press, Cambridge (2014). https://doi.org/10.1017/CBO9780511810114
69. Zarankin, T.G., Wall, J.A., Jr.: Negotiators' information sharing: the effects of opponent behavior and information about previous negotiators' performance. Negot. Confl. Manage. Res. **5**(2), 162–181 (2012)
70. Zeleznikow, J.: Using artificial intelligence to provide intelligent dispute resolution support. Group Decis. Negot. **30**(4), 789–812 (2021). https://doi.org/10.1007/s10726-021-09734-1
71. Zhang, S., Li, X., Zong, M., Zhu, X., Cheng, D.: Learning k for knn classification. ACM Trans. Intell. Syst. Technol. (TIST) **8**(3), 1–19 (2017)
72. Zhang, Y., et al.: Comparison of machine learning methods for stationary wavelet entropy-based multiple sclerosis detection: decision tree, k-nearest neighbors, and support vector machine. Simulation **92**(9), 861–871 (2016)
73. Zhang, Z.X., Han, Y.L.: The effects of reciprocation wariness on negotiation behavior and outcomes. Group Decis. Negot. **16**(6), 507–525 (2007)
74. Zhou, Z.-H.: Ensemble Methods: Foundations and Algorithms. Chapman and Hall/CRC, Boca Raton (2012). https://doi.org/10.1201/b12207

Preference Modeling and Multi-criteria Decision-Making

Ranking Potential Investors Using the FITradeoff Method and Value Focused Thinking in a Group Decision Problem

Geyse Maia da Silva, Eduarda Asfora Frej(✉), and Adiel Teixeira de Almeida

Universidade Federal de Pernambuco, UFPE, CDSID-Center for Decision Systems and Information Development, Av. Acadêmico Hélio Ramos, s/n, Cidade Universitária, 50.740-530, PE Recife, Brazil
eafrej@cdsid.org.br

Abstract. This work aims to address a problem of attracting companies for direct investments in the state of Pernambuco, Brazil, with the purpose to aid the Economic Development Agency of Pernambuco (ADEPE) to rank the potential investors. Considering different points of view of multiple stakeholders of ADEPE, the Value-Focused Thinking (VFT) and FITradeoff (Flexible and Interactive Tradeoff) techniques are applied in the construction of the multicriteria decision model. The problem structuring provides a better understanding of an investor's decision-making process and leads to a better pitch to persuade them to invest in Pernambuco. Furthermore, the structured problem can be analyzed by a quantitative method aiming to rank the best alternatives that could match with the proposal of the Pernambuco state as a win-win investment movement. For that, the multicriteria decision model is built based on the VFT approach and the FITradeoff method. A set of 25 alternatives were defined and evaluated with respect to six criteria. Throughout the process, the preference modeling is performed in an interactive and flexible way where the decision makers (DMs) can switch between elicitation by decomposition to holistic evaluations.

Keywords: Value-Focused Thinking (VFT) · Multicriteria Group Decision Making (MCGDM) · Preference Modeling · FITradeoff method

1 Introduction

The process of raising external resources to support the local economy is an important practice for any economy, especially in emerging economies. In the context of a marginal state of an emerging country as the state of Pernambuco, Brazil, the search for direct investments can be within the borders of the country. The proposed work is performed with the subject of investment attraction from companies that are already consolidated in Brazil and that have a potential need to expand their operations. The challenge is to anticipate the movement of the private sector and place the state of Pernambuco as a great place to invest.

Y. Maemura et al. (Eds.): GDN 2023, LNBIP 478, pp. 37–52, 2023.
https://doi.org/10.1007/978-3-031-33780-2_3

Defining the objectives that describe the potential of a company to expand operations and at the same time matches with the competitive advantage offered by the state is certainly not a trivial task. This work aims to structure and solve the problem for the active search for investments using the Value Focused Thinking (VFT) [1] approach and to evaluate the potential alternatives, based on a multicriteria decision model with the FITradeoff (Flexible and Interactive Tradeoff) method [2–4].

The use of a multiple criteria decision making/aiding (MCDM/A) is justified for its ability to assess multiple objectives and for having a rigorous formal structure [5, 6]. For public policies, the concept of bounded rationality of public decision makers was widely discussed [7]. However, the process and modeling of an MCDM/A problem fits the object of discussion, providing the appropriate elements to evaluate a decision that involves a public interest, by consistently and transparently applying its priorities in decision-making [8].

Several applications are found in the literature with the use of a multicriteria model within public organizations linked with investment attraction, such as: a multicriteria decision approach to determine regional investment strategies with the support of three multicriteria methods – TOPSIS, SAW and COPRAS [9], the decision support system using the multicriteria method Analytical Hierarchy Process (AHP) for the business location problem in London [10], location selection problem for sustainable landfill with AHP method [11]. Among other studies that use multicriteria methods for risk assessment [12], development of indexes [13] and organization of data [6] in the governmental scope.

In such complex decision problems involving multiple and conflicting criteria, the DM could be challenged to give precisely answers about his preferences. In that way, the FITradeoff method presents an interactive and flexible approach using partial information that provides less cognitive effort and has been used in different applications for the problematic of choice [4], ranking [3], sorting [14] and portfolio [15, 16]. The diversity of applications of the FITradeoff method allows its use in different areas of activity, such as: marketing [17], information technology [18–20], environmental and energy management [21, 22], industrial problems [23, 24]. The FITradeoff method combines two types of preference modeling in its structure: elicitation by decomposition and holistic evaluations [4], and therefore this method has the advantage of providing flexibility for the DM in decision process, to that the user can provide information in the way he/she feels more comfort with. Moreover, this method works based on partial information, in such a way that the elicitation questions are based on strict preference statements, turning the elicitation process less cognitively demanding [2].

The structure of this work is based on a framework for building multicriteria decision models developed by [5], which is divided in three main phases: preliminary phase, preference modeling and finalization. In the first phase, the problem structuring was performed with the VFT approach [1] and conducted with three decision makers with the purpose to define the objectives, criteria, and alternatives of the multicriteria problem. VFT is a well-known Problem Structuring Method (PSM), which aids DMs to better visualize the problem and find out objectives that guide them in the direction of their main goals, in a structured manner [1]. In the second phase, the preference modeling step was conducted with the three DMs using the FITradeoff method for the ranking problematic. The last phase is the construction of a recommendation based on the results of the individual ranking order of each DM.

The contribution of this paper is twofold. First, analyzing through a methodological perspective, it shows how a well-known problem structing method, the Value Focused Thinking, can be applied jointly with a multicriteria decision method, complementing each other on the process of structuring and solving a multicriteria decision problem. Second, the paper brings a practical contribution to improve the decision process of ranking potential investors for the government of Pernambuco, structuring the decision process and bringing up factors that were not previously considered in former decisions.

This paper is structured as follows. Section 2 presents a contextualization and description of the decision problem, as well as the objectives and criteria definition based on VFT. Section 3 is devoted to describe the preference modeling phase with the FITraeoff method, conducted with three decision makers. Section 4 presents the results obtained and discussion. Finally, final remarks are made in Sect. 5.

2 Problem Structuring Using Value Focused Thinking – VFT

This section aims to present the problem structuring process of the decision problem, which was conducted using the Value Focused Thinking (VFT) proposed by [1]. The VFT methodology helps to identify the fundamental and means objectives, and thus delve deeper into the statements of objectives, criteria and set of alternatives.

2.1 Contextualization of the Decision Problem and Actors

The object of study is grounded in the decision-making process of the Economic Development Agency of Pernambuco (ADEPE) regarding the need to characterize criteria that identify companies that have the potential to execute direct investments for the northeast region of Brazil, and that would be attractive to the state of Pernambuco in its strategic objective of economic development of the state.

ADEPE is an indirect entity administered by the government of Pernambuco, linked to the state's Secretariat for Economic Development (SDEC). Its objective is to support the economic and social development of the state of Pernambuco through actions that encourage and support the industrial, agro-industrial, commercial, service and handicraft segments with a focus on innovation.

The Agency has in its structure a specific sector for attracting investments, which, among its attributions, facilitates business by supporting potential investors in their dialogue with public bodies, with the aim of facilitating the decision-making process and implementation of new ventures. For that reason, it has a group of investment managers that search for potential investors and create a prospecting pool of enterprises that should guide them to invest resources for the action of persuading those investors to came to Pernambuco.

The greatest difficulty in the active search for investments is obtaining a set of companies whose criteria demonstrate potential interest in expanding their business and, at the same time, consider the Northeast a market opportunity. There are no defined processes for ranking potential ventures, leaving each investment manager to decide which companies to approach.

In this context, the prospecting process is essentially important, as it must seek among all the possibilities of companies on the market those that are most able to close a highly complex negotiation. Through active prospecting for new investments, the state of Pernambuco can position itself as a major player in the market.

Three decision makers were considered in this decision process: the direct leader of ADEPE and two other investment managers responsible for prioritizing companies that may be in the pool of actions to attract investments. In short, there are three decision makers that will be called DM1, DM2, DM3. In a governmental scope, there may be influences coming from the SDEC and the State Government directly. There are internal stakeholders as other state departments, city halls, the population, and local companies. And external stakeholders such as the government entities from other states and the federal entity that works to attract investments for Brazil.

2.2 Identification of Objectives

The process of identifying objectives requires creativity from the facilitator to extract the maximum amount of information from the decision maker and the actors involved in order to list the objectives that are of interest in the context of the decision. According to Keeney [1], in a problem with multiple stakeholders, the values should be structured separately for each one.

Based on the instruments proposed by [1], a semi-structured form was set up to guide the interview process. The three decision makers were interviewed individually. In a context of more than one interviewed decision maker, [25] suggests that each interviewee should have their lists of objectives and relationships separated until they are corrected and validated, to then be aggregated.

After collecting the data from the interview, a list of potential objectives was generated for each one of the DMs. The evaluation for fundamental objectives was addressed by the question "Why it's important?" [1]. If it leads to another objective, it's a mean to a fundamental objective. If there is no more answer for that question, it is a potential fundamental objective.

Table 1. Results from the data collected from the interview

Interviewed	Time duration	Potential fundamental objectives
DM1	00:56:00	10
DM2	00:48:42	10
DM3	00:36:45	14

Defining the key objectives is not an exact process, different perspectives could end up with completely distinguish hierarchy of fundamental objectives. With the aim of obtaining such objectives, interviews were conducted with the three DMs. Table 1 shows the interview details with each DMs: time duration of the interview, in the first column; and number of potential objectives found in the end of the process in the second column. The potential fundamental objectives obtained during the interviews were combined and structured for further analysis of the DMs. The objectives were evaluated in the context of the nine properties of the fundamental objectives proposed by [1]. The combined and structured fundamental objectives were presented to the DMs which validated in a consensual agreement a set of six fundamental objectives for the problem context.

Maximize the density of local production chains: Completing the production chain of an economic sector makes it possible for all materials and production inputs to be found in the state boundaries. It allows companies to reduce logistical costs. This conception contributes to the improvement of the economic scenario as a whole and the production chain becomes more competitive.

Maximize investments: This goal contributes to state revenue. The greater the investment, the greater the company's revenue, and the payment to the state.

Maximize job creation: This fundamental objective seeks investments that allow a high generation of employment. The state still values companies that are extensive in labor.

Minimize costs (logistics): This objective seeks to use the location of Pernambuco as a strategic differential. The fundamental objective highlights companies that already have customers in the region and that have a high freight cost for the distribution of their products. Decreasing this cost is good for local customers who will have the product without the high added transport cost and makes the state of Pernambuco economically viable for the investor.

Maximize brand positioning in the Northeast region (maximize sales): The fundamental objective is to value companies that do not yet have operations in the Northeast. These companies may be in the process of expansion planning and have not yet decided which state in the Northeast to settle in. These are companies that are consolidated and have a high degree of capillarity (they are present in more than one region).

Maximize the interiorization of development: the capacity to interiorate the development is a key objective to bring investments and job creation outside the metropolitan region. This objective searches for investments that can be taken to the interior of the state of Pernambuco.

2.3 Criteria Definition

The structuring of the objectives provides means for the use of quantitative methods which results in an in-depth understanding of the decision context. This process allows measuring the achievement of objectives through attributes, also called criteria. Keeney [1] defines an attribute as the degree to which a goal is achieved. Defining the criteria is the major concern at this stage of the problem because it is necessary the DM understand clearly what is being evaluated and how are the levels of achievement of each criterion.

Table 2 shows the six criteria for the problem discussed, all of them has a numerical scale associated. The type of criterion column indicates if the criterion is built on natural or constructed scale, or even if it is a proxy criterion [1]. The preference direction column indicates whether the criterion is to be maximized or minimized, according to the DMs' preferences.

Table 2. Definition of the criteria

Criteria	Definition	Type of criterion	Preference direction
C1-Sector priority	The priority levels of the economic sectors follow the strategic definition of the current government based on the density of local chains. The knowledge of the experts was used to segment the economic sectors into 4 levels, with the 1st highest priority and the 4th being the least priority	Constructed	Minimization
C2- Share capital	The share capital summarizes the company's accounting situation and investment capacity. The higher the capital, the greater the investment made for a new venture. The values are measured in R$ 1 MM (one million reals)	Natural	Maximization
C3 - Number of job positions	The number of jobs that the company has is an attribute to measure the company's job creation capacity. The higher the number of jobs, the greater the service to the objective	Natural	Maximization
C4- Distance (km)	It refers to the distance between the capitals and it is measured in KM. The longer the distance, the higher the cost to deliver the product to the market of Pernambuco and the greater the potential for cost reduction when settling in PE	Natural	Maximization

(continued)

Table 2. (*continued*)

Criteria	Definition	Type of criterion	Preference direction
C5- Capillarity	The capillarity of the company is constructed from the number of regions in which the company is present and the total number of states in which it has operation. The higher the capillarity, the greater the service to the objective. The constructed criterion has 7 levels	Constructed	Maximization
C6- MHDI	A proxy attribute was used for the fundamental objective of maximizing the interiorization. The Municipal Human Development Index (MHDI) of the place where the company has operations can be used as an approximation of how adherent the company is to settle in municipalities of less development. The lower the MHDI of the city of origin, the greater the tendency for it to settle in a region further away from the capital	Proxy	Minimization

2.4 Alternatives and Consequences Matrix

The generation of alternatives in a VFT approach is the most creative part of the decision process [1]. Structuring the objectives gives a clear idea of what are the means to achieve the goal. In the end, it is possible to speculate which are the best alternatives that should be evaluated in this decision process.

As a result of the means-end objectives network, a set of 25 enterprises were selected to be part of the set of alternatives. They were chosen based on the attributes that best represents the interest of the DMs.

The consequences matrix illustrated in Table 3 shows the evaluation of the alternatives in each criterion.

Table 3. Consequences matrix

Altern	C1	C2	C3	C4	C5	C6
A1	2	R$ 832.857,80	7059	2074	6	0,659
A2	2	R$ 92.040,12	13367	120	7	0,618
A3	1	R$ 82.633,21	2012	2061	2	0,689
A4	1	R$ 22.991,44	2446	2061	2	0,736
A5	1	R$ 68.030,94	1283	839	4	0,694
A6	1	R$ 1.734.659,21	661	800	2	0,658
A7	1	R$ 1.238.871,78	1998	2061	2	0,776
A8	2	R$ 11.804,14	2862	2061	2	0,73
A9	2	R$ 4.336,50	2477	120	4	0,7
A10	3	R$ 500.000,00	6346	2660	2	0,736
A11	3	R$ 237.018,86	3927	0	4	0,68
A12	1	R$ 101.105,25	1049	800	2	0,641
A13	1	R$ 22.582,55	1262	800	2	0,701
A14	1	R$ 277.417,78	656	2660	1	0,805
A15	4	R$ 178.879,19	785	3247	1	0,67
A16	4	R$ 263.495,02	807	839	1	0,677
A17	4	R$ 282.939,25	828	2660	1	0,764
A18	4	R$ 4.764,70	1760	2332	2	0,708
A19	2	R$ 155.759,39	3800	297	4	0,672
A20	2	R$ 21.150,61	5721	2338	2	0,693
A21	2	R$ 167.878,55	4870	2058	4	0,628
A22	2	R$ 72.805,32	4697	2061	2	0,695
A23	2	R$ 12.693,46	2289	2074	4	0,564
A24	3	R$ 143.335,50	2953	2332	4	0,755
A25	2	R$ 202.571,67	4219	2061	2	0,674

3 Preference Modeling with FITradeoff Method

The phase 2 of the framework proposed by [5] is the preference modeling in the multicriteria problem. The FITradoff method for the ranking problematic was applied to build an ordered list of the potential investors that should be prioritized, according to the company's objectives.

The Flexible and Interactive Tradeoff (FITradeoff) is a multicriteria method for eliciting the scale constants of the criteria (attributes) developed by [2], which aims to model the preferences of the decision maker using partial information into an additive model approach within a MAVT (Multi Attribute Value Theory) scope. The FITradeoff method combines two types of preference modeling in its structure [4]: elicitation by decomposition, in which the DM compares elements in the consequences space, similar to what happens in the classical tradeoff [26], but with partial information provided; and holistic evaluations, in which the DM compares elements in the alternatives space. The possibility of combination of these two distinct preference modeling approaches - switching between them at any time during the process, according to the DMs wishes - is a key flexibility feature of the FITradeoff method that enables the decision process to be conducted in a more efficient manner, with less information provided [4]. The information obtained by these two types of preference elicitation is converted into inequalities that act as constraints for a linear programming model that searches for dominance relations between alternatives, therefore constructing a ranking of them [3]. The FITradeoff method is operated by means of a Decision Support System (DSS), available at www.fitradeoff.org.

In this decision problem, all decision makers agreed on the values of each criterion for each alternative, in a sense that the consequences matrix in Table 3 was applied with all of them. The FITradeoff DSS was applied separately with each decision maker, guided by an analyst with well background on the method. The elicitation was conducted separately due to agenda limitation of the DMs.

Starting the preference modeling process with FITradeoff, the intracriteria value funcions were considered to be linear, for simplification purposes. However, in FITradeoff, it is also possible to consider nonlinear value function. Then, the decision maker ranks the criteria scaling constants (considering the ranges of consequences in each criterion). This process can be done by pairwise comparison, where the DM is asked about his preference for each pair of consequences, or by overall evaluation in FITradeoff DSS.

The result of this step of ranking of criteria scaling constants is shown in Table 4. The scaling constant of criterion c_i is denoted by k_{Ci} in Table 4. The ranking of each DM has significant differences. DM1 ranks in first position criterion C4, and for DM2 and DM3, the same criterion occupies the last position.

Table 4. Results of ordering criteria scale constants by DMs

Ranking	DM1	DM2	DM3
1°	k_{C4}	k_{C3}	k_{C3}
2°	k_{C1}	k_{C1}	k_{C2}
3°	k_{C5}	k_{C2}	k_{C6}
4°	k_{C2}	k_{C6}	k_{C5}
5°	k_{C6}	k_{C5}	k_{C1}
6°	k_{C3}	k_{C4}	k_{C4}

After the ranking of criteria scaling constants, the elicitation process is carried out in a flexible way, such that the DM can decide between elicitation by decomposition (Fig. 1) and holistic evaluations (Fig. 2), with the possibility of alternate between them. In the elicitation by decomposition, the DM compares two hypothetical consequences, which have the worst possible outcome in all criteria except for two of them (adjacent, according to the order presented in Table 4). The DM answers such questions by considering tradeoffs between these two criteria, choosing which consequence is preferred to him. By answering multiple questions of this type, this information is converted into inequalities for the criteria scaling constants, which form the so-called space of weights [2]. Those inequalities enter as constraint into a linear programming model that searches for dominance relations between alternatives, and then a (partial or complete) ranking is obtained, depending on the level of information obtained [3].

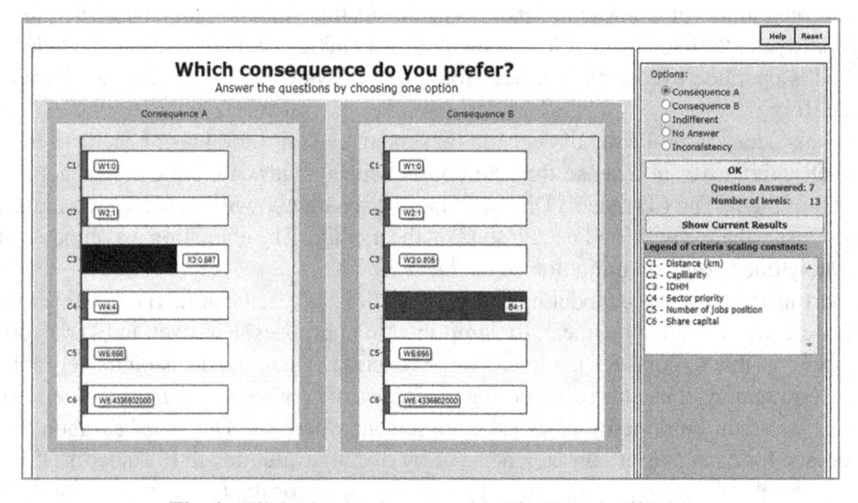

Fig. 1. Elicitation by decomposition in FITradeoff DSS

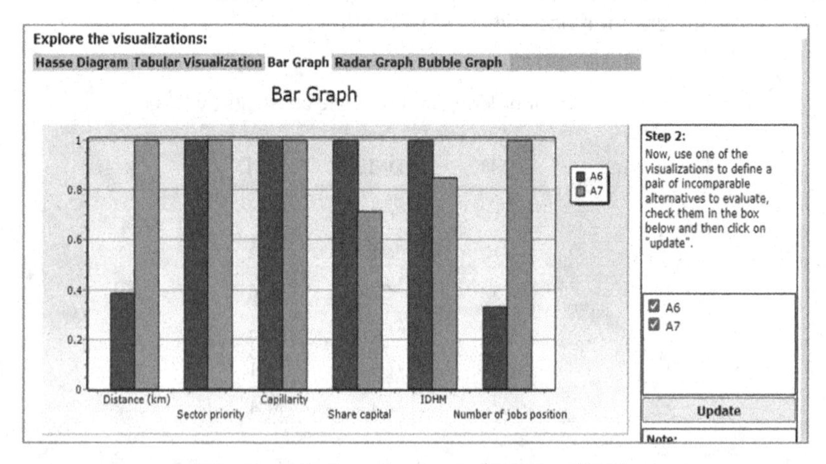

Fig. 2. Holistic evaluation in FITradeoff DSS

The other type of preference modeling consists on holistic evaluations, in which the DM compares alternatives directly, aided by graphical visualization (Fig. 2). In the ranking problematic, the holistic evaluation can be used to directly define a preference relation between two alternatives between which a dominance relation has not been defined yet. The FITradeoff DSS offers different types of graphical visualization to aid holistic analysis (bars, bubbles and radar graphics). Figure 2 shows a bar graphic that can be explored to compare two alternatives (A6, in blue; and A7, in orange) The graphic show the performance of these alternatives in each criterion, in a 0–1 ratio scale. In this graphic, criteria are ordered from left to right. By analyzing them in a comparative manner, the DM can decide which alternative is preferred, according to his preferences. It should be highlighted that the DM also has the option to switch back to the elicitation by decomposition, in case he does not feel confident (or does not want to) perform a holistic evaluation at that point [4]. Once a holistic judgment is performed, an inequality involving the global values of the two compared alternatives is obtained, and the space of weights is updated accordingly. It should be highlighted that the DM can alternate between the two types of preferences elicitation during the process, and he can also stop the elicitation process at any time, whenever he feels that the partial ranking obtained is enough for his purposes.

4 Results and Discussion

The elicitation process described in the previous section was carried out with the three decision makers, and the results are displayed in Table 5. For DM1, 11 questions were answered in the elicitation by decomposition and 03 holistic evaluations were made. As for DM2, 35 questions were answered in the elicitation by decomposition and no holistic evaluations were made. Finally, DM3 answered just 04 questions in the elicitation by decomposition and a complete pre-order has been found.

From Table 5, it can be seen that, in all three rankings, some alternatives remained tied in a certain ranking position. This is more evident in the ranking of DM2, which achieved 14 ranking positions, and the others achieved 20 ranking positions. It should be highlighted that the elicitation process in FITradeoff is flexible, in such a way that it can be interrupted whenever the DM is satisfied with partial results, not necessarily achieving a complete ranking, which was exactly was happened in this case.

As a prioritization problem, the ranking of the three DMs should be analyzed and compared. At first, it can be seen that the first two positions are the same for all DMs, with alternatives A1 and A2. Hence, it becomes clear that these two enterprises deserve to be the first one in which the state will prospect. Then, it is possible to analyze that the rankings of DM1 and DM3 are identical until the fifth position, and extremely similar in the following positions. In this sense, it is possible to see that the ranking of DM2 is a bit different from the other two.

In this case of divergence of rankings, some approaches could be considered to rank the results. One of the most widely used approaches to combine rankings is applying voting rules. The Borda procedure [27] is one of the most well-known voting rules, with which is possible to combine rankings by stablishing scores for alternatives according to the position that each alternative occupies in each individual ranking. By assuming that the DMs answered the FITradeoff elicitation question considering their actual preferences – without manipulation purposes -, the Borda rule could be suitable to combine

Table 5. Individual rankings

Ranking	DM1	DM2	DM3
1	[A1]	[A2]	[A1]
2	[A2]	[A1]	[A2]
3	[A21, A23]	[A20]	[A21, A23]
4	[A7]	[A10]	[A7]
5	[A6]	[A21]	[A6]
6	[A3]	[A22]	[A10, A20]
7	[A20]	[A25]	[A3]
8	[A4, A25]	[A19]	[A25]
9	[A5, A10][A10, A22][A22, A14]	[A4][A7][A3]	[A22]
10	[A24]	[A11]	[A4]
11	[A12]	[A6][A8][A9][A23][A5][A13][A12][A24]	[A24]
12	[A8]	[A14]	[A14]
13	[A19]	[A18]	[A5, A15][A15, A8]
14	[A13]	[A15, A16, A17]	[A12]
15	[A15]		[A19]
16	[A9]		[A13]
17	[A11]		[A18]
18	[A18]		[A9, A17]
19	[A17]		[A11]
20	[A16]		[A16]

these rankings. Table 6 shows the final ranking obtained combining the three rankings of Table 5 with Borda rule.

Hence, obtaining these results, ADEPE should follow the prioritization displayed in Table 6 to organize its potential investments. It should be highlighted, however, that other mechanisms could be used to achieve a final solution, such as other voting procedures [28], decision rules or even a consensual approach guided by a facilitator. Nevertheless, when the number of alternatives increases, achieving a consensus based on process support becomes even more challenging. In this sense, an analytical solution to combine rankings, such as using the Borda rule, seems more suitable in this case.

Table 6. Final group ranking, applying Borda rule

Position	Alternative
1	A1
2	A2
3	A21
4	A20
5	A7
6	A23
7	A10
8	A6
9	A3
10	A25
11	A22
12	A4
13	A24
14	A5, A14
15	A8, A12, A19
16	A13
17	A15
18	A9
19	A11
20	A18
21	A17
22	A16

5 Final Remarks

Decision situations involving multiple criteria and multiple decision makers at the same time inherently involve some challenging tasks, such as: structuring the problem in a coherent manner, stablishing objectives and criteria relevant for the DMs, evaluate the set of alternatives according to these criteria and aggregating different decision makers' preferences. In this work, those challenges were addressed in the context of prioritization of potential investors in the state of Pernambuco, Brazil. 25 companies were considered as alternatives, which were evaluated with respect of 6 criteria. Preferences of three decision makers were considered. In order to stablish the objectives and criteria, the Value Focused Thinking approach was applied. Then, the preferences elicitation process was conducted separately with each DM, using the FITradeoff DSS. The flexibility features of the FITradeoff method were explored during the process, in which the DMs could alternate between elicitation by decomposition and holistic evaluations. After achieving the three individual rankings, since different rankings were obtained, the Borda rule was applied to combine then and therefore obtain the final ranking of investors. Since the FITradeoff method gives the individual results for each DM, the Borda rule was applied as an alternative approach to aggregate individual rankings obtained by FITradeoff, in a simpler manner. However, other approaches could have been applied, such as a final meeting with all three DMs and a facilitator, in which one could try to show the others their point of views within a participative perspective, and therefore a consensus could be achieved. Or even other voting procedures, decision rules and aggregation techniques could have been performed.

Regarding the number of questions that each DM had to answer in the application of the FITradeoff method, it may vary depending on several aspects of the problem [29], such as number of criteria, alternatives topology and distribution of weights. However, it should be highlighted that the questions made in FITradeoff are strict preference questions, which are cognitively easier compared to other methods such as the classical tradeoff procedure [26], in which indifferent statements should be provided. Moreover, the FITradeoff method distinguishes from other MCDM methods by the advantage of combining to types of preference modeling (elicitation by decomposition and holistic evaluations), and giving the DM the flexibility to choose when to use each of them, according to the way in which he/she feels more comfortable with.

The perception of the decision makers for the structuring of the problem was positive, they agreed with the objectives and criteria presented. During the preference modeling process, the DMs showed good understanding of the questions and how they could indicate where to continue the elicitation, switch for a holistic evaluation or stop the process. At first, they preferred the elicitation by decomposition, but when the number of questions were rising there were some hesitations on the answers, and that was the moment in which the holistic evaluation was considered.

The next steps for implementation of the decision are to allocate the alternatives that are ranked in the first positions to the investment managers to work with specific strategies of investment attraction, that can include a propose of a set of opportunities and benefits the state could offer to the company.

Finally, it can be highlighted that the innovation and significance of this paper relies on both methodological and practical perspectives. Methodologically speaking a well-known PSM, the Value Focused Thinking, was applied jointly with the FITradeoff method, aiding on the objectives structuring phase. On a practical perspective, the decision process of ranking potential investors for the government of Pernambuco was improved with the use of structured methodologies to guide it and help the DMs, which could better think about the problem and take into account factors that were not previously considered in previous decisions.

For future studies, it would be interesting to evaluate the public investments allocated to the attraction of a new investments and consider the actual return of the new venture to the state.

Acknowledgments. The authors are most grateful for CNPq, FACEPE and CAPES, for the financial support provided.

References

1. Keeney, R.: Value Focused Thinking: A Path to Creative Decision Making. Harvard University Press, Cambrige (1992)
2. De Almeida, A.T., Almeida, J.A., Costa, A.P.C.S., Almeida-Filho, A.T.: A new method for elicitation of criteria weights in additive models: flexible and interactive tradeoff. Eur. J. Oper. Res. **250**(1), 179–191 (2016)
3. Frej, E.A., de Almeida, A.T., Costa, A.P.C.S.: Using data visualization for ranking alternatives with partial information and interactive tradeoff elicitation. Oper. Res. Int. J. **19**(4), 909–931 (2019). https://doi.org/10.1007/s12351-018-00444-2
4. de Almeida, A.T., Frej, E.A., Roselli, L.R.P.: Combining holistic and decomposition paradigms in preference modeling with the flexibility of FITradeoff. CEJOR **29**(1), 7–47 (2021). https://doi.org/10.1007/s10100-020-00728-z
5. DeAlmeida, A.T., et al.: Multicriteria and Multiobjective Models for Risk, Reliability and Maintenance Decision Analysis. ISORMS, vol. 231. Springer, Cham (2015). https://doi.org/10.1007/978-3-319-17969-8
6. Zackiewicz, M., Albuquerque, R., Salles Filho, S.: Multicriteria analysis for the selection of priorities in the Brazilian program of technological prospection – Prospectar. Innovation **7**(2–3), 336–348 (2005). https://doi.org/10.5172/impp.2005.7.2-3.336
7. Simon, H.A.: Administrative Behavior: A Study of Decision-Making Processes in Administrative Organization, 2nd edn. Macmillan, New York (1957)
8. Kurth, M.H., Larkin, S., Keisler, J.M., Linkov, I.: Trends and applications of multi-criteria decision analysis: use in government agencies. Environ. Syst. Decis. **37**(2), 134–143 (2017). https://doi.org/10.1007/s10669-017-9644-7
9. Ustinovichius, L., Komarovska, A., Komarovski, R.: Methods of determining the region's investment strategy. Proc. Eng. **182**, 732–738 (2017). https://doi.org/10.1016/j.proeng.2017.03.190
10. Weber, P., Chapman, D.: Location intelligence: an innovative approach to business location decision-making. Trans. GIS **15**(3), 309–328 (2011). https://doi.org/10.1111/j.1467-9671.2011.01253.x
11. Aksoy, E., San, B.T.: Geographical information systems (GIS) and multi-criteria decision analysis (MCDA) integration for sustainable landfill site selection considering dynamic data source. Bull. Eng. Geol. Env. **78**(2), 779–791 (2017). https://doi.org/10.1007/s10064-017-1135-z
12. Hussain, J., Zhou, K., Guo, S., Khan, A.: Investment risk and natural resource potential in "Belt & Road Initiative" countries: a multi-criteria decision-making approach. Sci. Total Environ. **723**, 13 (2020). https://doi.org/10.1016/j.scitotenv.2020.137981
13. Silva, M.C., Costa, H.G., Gomes, C.F.S.: Multicriteria decision choices for investment in innovative upper-middle income countries. Innov. Manage. Rev. **17**(3), 321–347 (2020)
14. Kang, T.H.A., Frej, E.A., de Almeida, A.T.: Flexible and interactive tradeoff elicitation for multicriteria sorting problems. Asia Pac. J. Oper. Res. **37**, 2050020 (2020)
15. Frej, E.A., Ekel, P., de Almeida, A.T.: A benefit-to-cost ratio based approach for portfolio selection under multiple criteria with incomplete preference information. Inf Sci **545**, 487–498 (2021)
16. Marques, A.C., Frej, E.A., de Almeida, A.T.: Multicriteria decision support for project portfolio selection with the FITradeoff method. Omega **111**, 102661 (2022)
17. Shukla, S., Dubey, A.: Celebrity selection in social media ecosystems: a flexible and interactive framework. J. Res. Interact. Market. **16**(2), 189–220 (2021). https://doi.org/10.1108/JRIM-04-2020-0074

18. Henriques, A.P., de Gusmão, C., Medeiros, P.: A model for selecting a strategic information system using the FITradeoff. Math. Probl. Eng. **2016**, 1–7 (2016). https://doi.org/10.1155/2016/7850960
19. AlvarezCarrillo, P.A., Roselli, L.R.P., Frej, E.A., de Almeida, A.T.: Selecting an agricultural technology package based on the flexible and interactive tradeoff method. Ann. Oper. Res. **2018**, 1–16 (2018). https://doi.org/10.1007/s10479-018-3020-y
20. Poleto, T., Clemente, T.R.N., de Gusmão, A.P.H., Silva, M.M., Costa, A.P.C.S.: Integrating value-focused thinking and FITradeoff to support information technology outsourcing decisions. Manag. Decis. **58**(11), 2279–2304 (2020). https://doi.org/10.1108/md-09-2019-1293
21. Fossile, D.K., Frej, E.A., da Costa, S.E.G., de Lima, E.P., de Almeida, A.T.: Selecting the most viable renewable energy source for Brazilian ports using the FITradeoff method. J. Clean. Prod. **260**, 121107 (2020)
22. Kang, T.H.A., Soares, A.M.C., Jr., De Almeida, A.T.: Evaluating electric power generation technologies: a multicriteria analysis based on the FITradeoff method. Energy **165**, 10–20 (2018)
23. Frej, E.A., Roselli, L.R.P., de Almeida, J.A., de Almeida, A.T.: A multicriteria decision model for supplier selection in a food industry based on FITradeoff method. Math. Probl. Eng. **2017**, 1–9 (2017)
24. Pergher, I., Frej, E.A., Roselli, L.R.P., de Almeida, A.T.: Integrating simulation and FITradeoff method for scheduling rules selection in job-shop production systems. Int. J. Prod. Econ. **227**, 107669 (2020)
25. Kunz, R.E., Siebert, J., Mutterlein, J.: Combining value-focused thinking and balanced scorecard to improve decision-making in strategic management. J. Multi-Crit. Decis. Anal. **23**, 225–241 (2016)
26. Keeey, R.L., Raiffa, H.: Decision Making with Multipleobjectives, Preferences, and Value Tradeoffs. Wiley, New York (1976)
27. de Borda, J.C.: Mémoire sur les élections au scrutin, Mémoire de l'Académie Royale. Histoire de l'Académie des Sciences, Paris, pp 657–665 (1781)
28. Teixeira, A., de Almeida, D., Morais, C., Nurmi, H.: Systems, Procedures and Voting Rules in Context: A Primer for Voting Rule Selection. Springer International Publishing, Cham (2019). https://doi.org/10.1007/978-3-030-30955-8
29. Mendes, J.A.J., Frej, E.A., Teixeira, A., de Almeida, J., de Almeida, A.: Evaluation of flexible and interactive tradeoff method based on numerical simulation experiments. Pesquisa Operacional **40**, 1–25 (2020). https://doi.org/10.1590/0101-7438.2020.040.00231191

Using Unfolding Analysis and MARS Approach for Generating a Scoring System from a Group Preference Information

Tomasz Wachowicz[1], Ewa Roszkowska[2](✉), and Marzena Filipowicz-Chomko[2]

[1] University of Economics in Katowice, 1 Maja 50, 40-287 Katowice, Poland
tomasz.wachowicz@uekat.pl

[2] Bialystok University of Technology, Wiejska 45A, 15-351 Bialystok, Poland
{e.roszkowska,m.filipowicz}@pb.edu.pl

Abstract. This paper proposes a new approach for determining the negotiation offer scoring system for a negotiator out of the group preference information. The proposed algorithm, unMAGIC (***un**folding and **MA**RS-based **G**roup preference **I**nformation **C**omputation*), is a hybrid of the unfolding and MARS (*Measuring Attractiveness near Reference Solution)* methods. In its first steps, the experts express their preferences over some exemplary negotiation offers, predefined according to the principles of the MARS approach, by declaring their rank order. Then, the rank orders are processed according to the principles of unfolding analysis to produce two-dimensional evaluations of each offer that describe how much it is preferred from the aggregated viewpoint. Offers and experts are also depicted on the plane, showing the distances to each other, which helps the negotiator better understand the variety of recommendations. Then, using the principles of MARS, these distances are used to build the additive scoring system for the negotiator. Finally, we verify how our hybrid approach works using a dataset from the prenegotiation experiment based on a bilateral negotiation case.

Keywords: Preference elicitation · Negotiation scoring system · Collective negotiation profile · Unfolding method · MARS method

1 Introduction

One of the negotiators' prenegotiation activities is structuring the negotiation problem and developing the negotiation template and scoring system [1, 2]. The former describes the negotiation issues and the feasible spaces of their resolution levels. The latter defines scores for all the elements of the negotiation template and is used by negotiation support systems or third parties to support the negotiators' decision and ensure efficient and fair compromises. The theory of negotiation analysis recommends multiple criteria decision-aiding (MCDA) methods to help negotiators solve this problem [3–6]. In typical negotiations, for each issue, the feasible resolution levels (options) are predefined, and offers are determined as combinations of these options, which makes the potential decision matrix to be evaluated large.

Y. Maemura et al. (Eds.): GDN 2023, LNBIP 478, pp. 53–66, 2023.
https://doi.org/10.1007/978-3-031-33780-2_4

The scale of the problem and the negotiators' behavioral limitations may make analyzing preferences and determining the scoring system in prenegotiation difficult and prone to many errors [7, 8]. Therefore various techniques have been developed to help negotiators in prenegotiation decision analyses [1, 9, 10]. Most focus on exploring the negotiators' preferences with tools tailored to the negotiation problem's specificity or the parties' cognitive capabilities [11, 12]. Another approach is to provide the negotiator with information regarding preferences typical for the group of other negotiators to which she/he is similar (concerning, e.g., some business characteristics). With data describing the preferences of a broad population of negotiators, it is possible to construct models of collective preferences for different groups of negotiators [13, 14]. Such solutions are also used in practice, for example, in electronic marketplaces designed for organizing multi-issue auctions or tenders. There, the predefined scoring functions are suggested to auction organizers deriving from the ones earlier used by similar types of customers. One such example is the idea used in organizing an e-marketplace named OpenNexus[1], which is used for contracting by many public institutions in Poland.

Since the number of negotiators clustered in a particular group of similarity may be large, the analytical approach used to determine a collective profile implements the techniques from the data analysis rather than group decision-aiding. For instance, Piasecki et al. [14] proposed an approach that used a fuzzy representation of a collective preference profile through the trapezoidal fuzzy numbers constructed out of the preferences of individual negotiators. Such an approach allowed representing the collective preferences with all the nuances resulting from the individual declarations. However, the resulting global fuzzy scores of offers did not allow for their easy comparison.

Therefore, in this paper, we propose a new approach for determining the negotiation offer scoring system out of the group preference information that hybridizes the mechanics of the unfolding analysis [15] and the MARS (*Measuring Attractiveness near Reference Solution*) technique [10]. In this approach, we assume that the group of experts expresses their preferences over selected negotiation offers predefined according to the principles of MARS by declaring their rank order. Then, the rank orders are processed according to unfolding analysis to produce two-dimensional evaluations of each offer that describe how much preferred it is from the collective (aggregated) viewpoint. Then, using MARS again, we decompose these evaluations to find the scores of all options in the template and build additive scoring systems for the negotiator.

There are some objective reasons for hybridizing these two approaches. First, the MARS technique cannot produce the group decision recommendation alone, as it was designed to support a single decision maker (DM) only. The unfolding method allows nicely visualizing of the opinion of a group of experts over a predefined set of alternatives. Unfortunately, it does not give recommendations on how this set should look nor the procedure for determining the disaggregated scoring system out of this set.

The hybridized approach has some advantages. First, incorporating MARS principles to predefine the set of exemplary offers allows for building a set of alternatives that are easy to compare. Their evaluation requires, at most, a consideration of a single trade-off between two negotiation issues. Unfolding allows describing alternatives, previously ranked by experts, through the coordinates on the Euclidean space, which can

[1] https://platformazakupowa.pl/

be later used to produce the cardinal scores from the MARS algorithm. Consequently, the approach reduces the cognitive requirements imposed on experts.

The paper consists of three more sections. Section 2 presents the theoretical issues related to defining the negotiation template and building the scoring system using the proposed hybridized approach that mixes unfolding analysis and the MARS method. Section 3 describes the prenegotiation experiment that provides the data for the analyses using our hybridized approach and shows how it works using experimental data. We summarize the approach in conclusions.

2 Methodology

2.1 Negotiation Template and Scoring System for Negotiation Support

In negotiation analysis, notions of negotiation template and scoring system are key-important from the viewpoint of any negotiation support that can be offered to the parties [16]. The template defines the structure of the negotiation problem, while the scoring system quantitatively describes the negotiators' preferences for the template's elements[2]. Let $G = \{g_i\}_{i=1,\ldots,m}$ denotes a set of m issues to be discussed in negotiations, and $X_i = \left\{x_{i,j}\right\}_{j=1,\ldots,n_i}$ is a set of n_i salient options (potential resolution levels) for the issue g_i[3]. The negotiation template can be defined as $m+1$-tuple

$$T = \left\{G, \{X_i\}_{i=1,\ldots,m}\right\}. \tag{1}$$

Consequently, the scoring system is a corresponding $m+1$-tuple

$$S = \left\{w, \{V_i\}_{i=1,\ldots,m}\right\}, \tag{2}$$

where $w = [w_1, \ldots w_m]$ is a vector of issue weights such that $\sum_{i=1}^{m} w_i = 1$, and $V_i = \left\{v_{i,j}\right\}_{j=1,\ldots,n_i}$ is a set of cardinal scores, usually scaled to [0;1]-range, describing the attractiveness of options $x_{i,j}$. With the negotiation problem defined by template T and scoring system S, any negotiation offer a from a feasible negotiation space $A = \prod_{i=1,\ldots,m} X_i$ can be evaluated using a simple additive scoring formula

$$V(a) = \sum_{i=1}^{m} \sum_{j=1}^{n_i} w_i v_{i,j} z_{i,j}^{a}, \tag{3}$$

where $z_{i,j}^{a}$ is a binary variable indicating whether the option $x_{i,j}$ builds offer a.

The fact that offers can be quantitatively evaluated thanks to an adequately defined scoring system S opens many possibilities for offering asymmetric and symmetric negotiation support and designing negotiation support systems [2, 16–18].

---

[2] Classically, it is assumend that additive and fully compensatory model of preferences may be applied to adequately descibe negotiator's preferences.

[3] For a discussion on how such a definition may describe continuous negotiation problems, see [2]

2.2 The Multidimensional Scaling and Unfolding Methods – a Short Overview

Assuming the negotiator is supported in determining the scoring system S by the experts, an issue of adequately determining a single recommendation that represents a common viewpoint of all these experts arises. If the number of experts is not high, some notions of group decision-making may be applied [19]. Otherwise, e.g. when crowdsourcing or the ideas of knowledge society are applied, the notion of consensus is senseless. In that case, other tools from data analysis may be more efficient, or some generalizations of classic decision-making approaches need to be developed [20, 21].

One of the potential techniques to be used in such a situation is unfolding analysis that derives from multidimensional scaling (MS) [22]. MS is the process of finding the configuration of points (representing objects under consideration) in a space with a given number of dimensions. The distances between these points should represent the dissimilarities between objects in the most accurate way. The starting point of the MS procedure is the coordinates of objects (data points) in a space with a large number of dimensions and the matrix of distances (dissimilarities) or similarities between individual objects. The unfolding method [15] is one of the MS methods used in preference representation. It aims to detect a common space of points representing respondents and examined objects and assess the relationships between objects and respondents based on the configuration of the received points. In unfolding, input data are denoted by a preference matrix, where rows represent respondents and columns represent objects. Searching for coordinates is iterative and is completed when the value of a defined criterion function meets an assumed threshold or if the previously assumed number of iterations has been reached. Generally, the unfolding, aimed at scaling into, e.g., two-dimensional space, consists of searching for the coordinates of points in geometric space that optimally reproduces distances in the input data matrix. The distances of objects from the centroid of coordinates may reflect their collective attractiveness from the viewpoint of all respondents.

2.3 Holistic Declaration of Preferences in Multi-issue Negotiation – MARS Method

Many MCDA techniques may be used to determine the scoring system as required by formula (2) (see [23]). However, they differ regarding how the preference information is derived from DM. One group of methods assumes DMs can express their preferences directly on the disaggregated level, i.e., they explicitly define the ratings $v_{i,j}$ or the marginal value functions that could be used to determine them. Another one assumes that those marginal value functions or $v_{i,j}$ ratings can be derived from the preferences defined at the holistic level for the examples of alternatives (offers). In other words, instead of declaring ratings for elements of template T, the selected alternatives from A are considered by DMs, who may declare the order of them, assign them to the predefined quality categories, or assign them global numerical ratings. The holistic approach has been suggested earlier as a negotiation support tool [2, 9, 24, 25].

One of these holistic methods is MARS which has been used as a stand-alone decision-aiding tool for negotiation support or as a facilitating technique for other holistic approaches [2, 10]. MARS is a hybrid of ZAPROS [26] and MACBETH [27] methods.

Its big advantage is the precise recommendation of the set of exemplary offers that use all the options defined in the template T. When the ideal offer is considered as a reference solution a_I (consisting of most preffered options for all issues), all remaining ones that build the set of reference alternatives (H_{nRIS}) differ from a_I in resolution level of one issue only. Consequently, any two offers from H_{nRIS} differ on resolution levels of at most two issues. Hence, if negotiator compares them she must only consider which concession (on issue g_k or g_l) is most profitable to her. It is cognitively far more easier than comparing offers consisting of entirely different options.

Another characteristic of MARS is that it produces the joint cardinal scale out of the holistic evaluations of offers, which is based on the notion of distance of a particular offer to the reference one (a_I). These distance values are directly related to the poor performance of the non-ideal option that builds a particular offer from H_{nRIS}. Therefore, assuming that the marginal scoring functions are scaled to a particular range – they may be easily decoded and produce the entire scoring system S (i.e., weights and option ratings). The fact that both unfolding and MARS operate with notions of distances makes them formally legitimate to be hybridized to produce the scoring system out of the recommendations of experts or the participants of the crowdsourcing process. We will propose such a hybrid algorithm for determining the scoring system out of holistically defined preferences over the set of predefined negotiation offers in Sect. 2.5.

2.4 Analyzing the Coherence of Information Provided by the Group – Kendall's Coefficient of Concordance

Before the aforementioned analysis is performed, one should check how coherent the individual recommendations provided by subsequent experts are. If these recommendations differ too much, there may be no pattern in experts' preferences that we should suggest to the negotiator as representing the collective preferences well. Therefore, an adequate checkup procedure should be suggested to identify such a problem.

Assume that there is $p = 1, 2, \ldots, P$ experts that evaluate $q = 1, 2, \ldots, Q$ negotiation offers by assigning them ranks from 1 to Q, without ties. Let $r_{q,p}$ be a rank that pth expert assigned to qth offer, $R_q = \sum_{p=1}^{P} r_{q,p}$ be a total rank given to qth offer, $\overline{R}$ be the mean value of total ranks R_q ($q = 1, 2, \ldots, Q$), and SQD be the squared deviation given by the formula $SQD = \sum_{q=1}^{Q} \left(R_q - \overline{R}\right)^2$. Kendall's coefficient of concordance (or Kendall's W coefficient) of experts' opinions is defined as:

$$W = \frac{12SQD}{P^2\left(Q^3 - Q\right)}. \tag{4}$$

The concordance coefficient $W \in [0, 1]$. The higher the values W, the greater consistency of rankings among experts. If $W = 1$, then we have a complete agreement of individual recommendations, i.e., all experts obtained the same order of offers. If $W = 0$, then the experts' rankings may be regarded as random. According to [28], we may interpret W values using the following rule of thumb: $W \in [0; 0.2)$ – slight agreement, $W \in [0.2; 0.4)$– fair agreement, $W \in [0.4; 0.6)$– moderate agreement, $W \in [0.6; 0.8)$– substantial agreement and $W \in [0.8; 1)$ almost perfect agreement. Let us note that in the case of ties, the correction factors should be computed in W.

2.5 A Hybrid Approach to Determine the Scoring System from Collective Preference Declarations

With all the above methodological considerations, we will propose below an unMAGIC algorithm that hybridizes the unfolding and MARS methods to produce the negotiation scoring system out of the recommendations of a group of experts. This algorithm consists of the following steps:

- **Step 1.** Define the negotiation template T for the negotiation problem under consideration.
- **Step 2.** Build the set of reference alternatives H_{nRIS} using MARS principles.
  H_{nRIS} should be comprised of all feasible alternatives from A that consist of the best resolution levels for all criteria but one together with the alternative, which is the ideal solution. For a discretely defined negotiation template, the cardinality of such a set is equal to $|H_{nRIS}| = 1 + \sum_{i=1}^{m}(|X_i| - 1)$.
- **Step 3.** Collect individual recommendations of experts regarding the rank order of alternatives from H_{nRIS}.
- **Step. 4.** Verify the concordance of recommendations by determining Kendall's W coefficient for the dataset of recommendations gathered in step 3.
- **Step 5.** Perform two-dimensional unfolding analysis:

  - identify potential outlier recommendations (e.g., from the unfolding plane), and if the assumed minimum concordance threshold of W is not met, remove them and go to step 4,
  - for each alternative, determine its Euclidean distance from the point representing the opinion of the group (x_o, y_o) using the coordinates determined from the unfolding analysis:

$$L_q = \sqrt{(x_q - x_o)^2 + (y_q - y_o)^2} \tag{5}$$

  where x_q (y_q) are the coordinates of qth offer; x_0 (y_o) are the coordinates of the point representing the expert group's opinion (e.g., centroid point).

- **Step 6.** Normalize the distances L_q using the notion of min-max normalization, i.e.

$$\tilde{L}_q = \frac{L_q - \max L_q}{\min L_q - \max_{L_q}}. \tag{6}$$

- **Step 7.** Assign ratings to options using MARS principles:

  - Assign a rating equal to 1 to the best option within each X_i.
  - To each non-best option, assign a rating equal to the normalized distance of the offer this non-best option comprises.

- **Step 8.** Determine the global ratings $V\left(a_q\right)$ for all offers from H_{nRIS} and worst offer a_{AI} that consists of the worst options for each issue using formula (3).
  When determining the global scores, weights $w_i = 1$ for each issue $i = 1, 2, \ldots, m$ should be temporarily assumed.

- **Step 9.** Determine the rescaled global ratings $\tilde{V}(a_q)$ from $V(a_q)$ computed for the set $H_{nRIS} \cup a_{AI}$ using min-max normalization.
- **Step 10.** Compute [0; 1]-normalized scoring system for negotiation offers:
  - Determine issue weights using the following rule

$$w_i = 1 - \tilde{V}(a_q) \text{ if the worst option of th issue builds offer } a_q \quad (7)$$

  - Rescale the option ratings obtained in step 7 using min-max normalization within each issue separately.

3 Application Algorithm for the Determination of Negotiation Scoring System Based on Group Preference Information

To verify how the proposed hybrid algorithm works, we use a dataset from the prenegotiation experiment based on a bilateral negotiation case implemented in the Inspire system [17]. In our experiment, we asked the participants to holistically define their preferences over exemplary negotiation offers built out of the template defining this negotiation case. They were instructed that the rank orders they built would be used to determine the scoring system for the template. The participants were 98 students (58% of females; 42% of males) from three Polish Universities, and this experiment was a part of their computer science, economics, logistics, mathematics, and international relations courses. They all took prior training regarding negotiation support, analysis, and preference declarations (both in holistic and disaggregated ways) for determining the negotiation offer scoring systems (e.g., simple direct rating or UTA techniques). The rank orders of exemplary offers the participants declared during the experiment allow us to illustrate below our scoring mechanism.

- **Step 1.** The Inspire negotiation problem concerned designing a contract for a new rising star singer by her agent (Fado) and the representative of the music company (Mosico). In our experiment, all participants played the role of Mosico agents. They were asked to consider the negotiation template T defined through four issues and the sets of predefined feasible options, for which the principal's preferences were described verbally and visualized graphically for each negotiation party. The graphical representation of the template and preference information for the Mosico party is shown in Fig. 1.
- **Step 2.** The set of thirteen reference offers H_{nRIS} was built. The offers (shown in Table 1) are constructed with the best options (bolded) for all issues but one and the ideal solution which consists of all best options for all issues (offer O8). For details on generating H_{nRIS} set see [10].
- **Step 3.** We used the dataset from the experiment to collect individual recommendations from respondents regarding the rank order of offers from H_{nRIS}. The experimental setup allowed respondents to declare only the total rank order of offers, i.e., no two offers could be considered equally attractive (no rank ties occurred). However, it must be noted that the approach generally allows the experts to define the same ranks (ties). The summary of rank declarations by respondents is shown in Table 2.

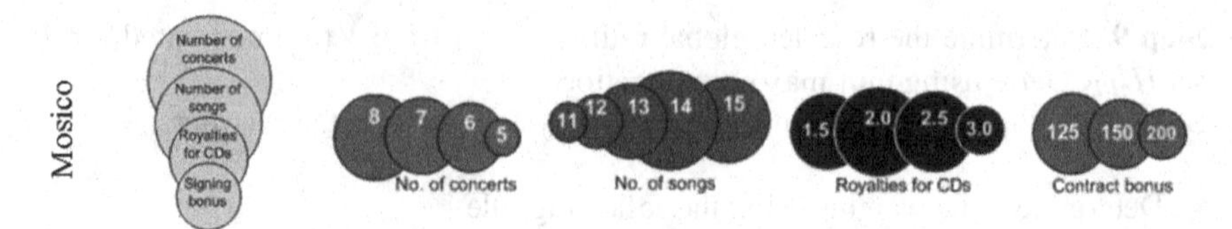

Fig. 1. Visual representation of template and Mosico preferences in Inspire negotiation

Table 1. The offers from the set H_{nRIS}

Offer id	Resolution levels building an offer
O1	5 concerts, **14 songs, 2% royalties, 125,000$ contract**
O2	6 concerts, **14 songs, 2% royalties, 125,000$ contract**
O3	7 concerts, **14 songs, 2% royalties, 125,000$ contract**
O4	**8 concerts,** 11 songs, **2% royalties, 125,000$ contract**
O5	**8 concerts,** 12 songs, **2% royalties, 125,000$ contract**
O6	**8 concerts,** 13 songs, **2% royalties, 125,000$ contract**
O7	**8 concerts, 14 songs,** 1.5% royalties, **125,000$ contract**
O8	**8 concerts, 14 songs, 2% royalties, 125,000$ contract**
O9	**8 concerts, 14 songs, 2% royalties,** 150,000$ contract
O10	**8 concerts, 14 songs, 2% royalties,** 200,000$ contract
O11	**8 concerts, 14 songs,** 2.5% royalties, **125,000$ contract**
O12	**8 concerts, 14 songs,** 3% royalties, **125,000$ contract**
O13	**8 concerts,** 15 songs, **2% royalties, 125,000$ contract**

- **Step 4**. Kendall's coefficient of concordance was determined, and an independence test was performed to verify the similarity of respondents' recommendations. We assumed the threshold value $W = 0.7$ as satisfactory for performing unfolding analysis. The degree of concordance for our respondents was $W = 0.745$, which can be interpreted as substantial agreement [28]. The chi-squared test confirms its significance at $p < 0.001$. Therefore, we will move to step 5 with the full dataset as an input for unfolding analysis.
- **Step 5.** Two-dimensional unfolding analysis was performed using SPSS 28 software. For the input data describing respondents' ranks (rows) assigned to negotiation offers (columns), the proximities in unfolding were set as dissimilarities and transformed by rows ordinally. The classical initial configuration with Spearman-based imputation and standard iteration criteria were used. Finally, after 171 iterations, the Stress function was obtained at 0.175 with a penalty component 2.083. The standard test benchmarking the Stress value to those from scaling random data confirms significant

Table 2. Distribution of ranks for offers from H_{nRIS} from participants declarations

Rank	O1	O2	O3	O4	O5	O6	O7	O8	O9	O10	O11	O12	O13
1	1	0	4	0	0	2	4	82	2	2	1	0	0
2	0	2	0	0	1	1	6	6	65	2	7	0	8
3	2	0	2	0	0	3	10	3	10	44	16	1	7
4	0	0	1	0	1	2	6	2	8	18	44	6	10
5	0	1	5	2	0	0	29	1	7	6	15	13	19
6	0	4	1	0	0	3	22	0	1	12	9	38	8
7	3	1	7	1	1	12	6	2	2	5	2	19	37
8	0	2	8	1	5	52	8	0	2	6	0	8	6
9	0	3	16	2	50	13	2	2	0	1	3	5	1
10	3	5	13	42	22	3	3	0	0	1	0	5	1
11	2	5	41	30	11	6	0	0	0	0	0	2	1
12	2	73	0	16	4	1	0	0	1	0	1	0	0
13	85	2	0	4	3	0	2	0	0	1	0	1	0
min	1	2	1	5	2	1	1	1	1	1	1	3	2
max	13	13	11	13	13	12	13	9	12	13	12	13	11
Mean rank	12.3	11.1	9.0	10.6	9.5	7.8	5.4	1.5	2.8	4.4	4.3	6.6	5.7
Dominant rank	13	12	11	10	9	8	5	1	2	3	4	6	7

differences at $p < 0.001$. The joint plot representing unfolding results is shown in Fig. 2.

The analysis of Fig. 2 allows us to determine the preferences of the group of respondents for negotiation offers. In general, the less distant the alternative is from the centroid of the points representing experts, the more it is preferred. Consequently, the most preferred offers are O8 and O9, and the last is O1. The joint graph shows, however, that O1 seemed more preferred than O8 and O9 by respondents 32, 33, and 51. However, these respondents differ in preference declarations from others, as a big cloud of blue points representing the majority of respondents lies far to the left. The detailed coordinates obtained from the unfolding analysis for offers and their distances from the centroid $(x_o, y_o) = (-0.462, 0.415)$ representing the opinion of group experts are shown in Table 3 (columns 2–4).

- **Step 6**. The normalized distances $\tilde{L}_q$ (raw offer scores) are determined from formula (6) – see Table 3, column 5.
- **Step 7**. Using MARS, the options are associated with each score $\tilde{L}_q$ (Table 3, column 6). For instance, offer O4 consists of the best options for all remaining issues but suggest 11 songs be recorded (while the best option is 14). Therefore its distance

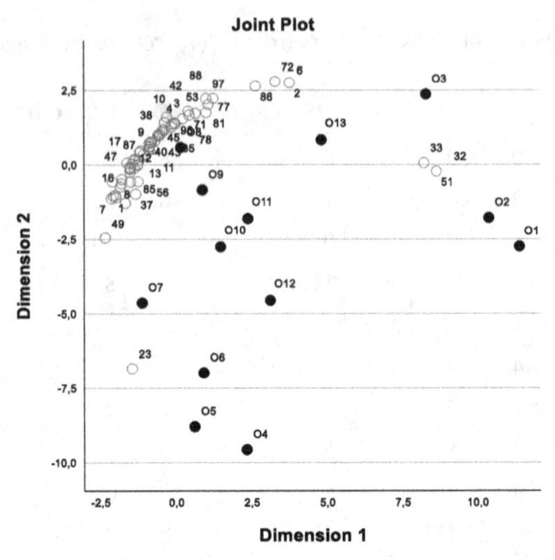

Fig. 2. Configuration of points representing negotiation offers from H_{nRIS} and respondents obtained by unfolding analysis

from the coordinate of the centroid, $L_q \approx 10.360$, and the consequent raw score ($\tilde{L}_q$=0.162) is linked to the poor evaluation of the attractiveness of option 11.

Table 3. The coordinates of points representing negotiation offers, Euclidean distances, normalized distances, and associated options.

Offer	Coordinate1	Coordinate 2	Euclidean distance L_q	Normalized distances $\tilde{L}_q$	Associated option(s)
O1	11.3645	– 2.7476	12.242	0.000	5
O2	10.3435	– 1.7799	11.025	0.105	6
O3	8.2464	2.3606	8.922	0.286	7
O4	2.3116	– 9.5666	10.360	0.162	11
O5	0.5873	– 8.7949	9.270	0.256	12
O6	0.8878	– 6.9831	7.520	0.406	13
O7	-1.1628	– 4.6361	5.100	0.614	1.5
O8	0.1283	0.5859	0.614	1.000	8, 14, 2, 125
O9	0.8427	– 0.8405	1.810	0.897	150
O10	1.4449	– 2.7587	3.703	0.734	200
O11	2.3535	– 1.8021	3.584	0.745	2.5
O12	3.0951	– 4.5552	6.112	0.527	3
O13	4.7792	0.8336	5.257	0.601	15

- **Step 8.** Determining the global ratings $V(a_q)$ for all offers from $H_{nRIS} \cup a_{AI}$ using normalized distances $\tilde{L}_q$ (Table 4, columns 2–6).
- **Step 9.** The rescaled global ratings $\tilde{V}(a_q)$ and the corresponding ranks are determined from $V(a_q)$. The results are presented in Table 4, columns 7–8.

Table 4. The rating offers from the set H_{nRIS} obtained by the MARS-based approach

Offer	Concert	Songs	Royalties	Contract	Global value $V(a_q)$	Rescaled Global Value $\tilde{V}(a_q)$	Rank
O1	0.000	1.000	1.000	1.000	3.000	0.612	13
O2	0.105	1.000	1.000	1.000	3.105	0.653	12
O3	0.286	1.000	1.000	1.000	3.286	0.723	9
O4	1.000	0.162	1.000	1.000	3.162	0.675	11
O5	1.000	0.256	1.000	1.000	3.256	0.711	10
O6	1.000	0.406	1.000	1.000	3.406	0.769	8
O7	1.000	1.000	0.614	1.000	3.614	0.850	5
O8	1.000	1.000	1.000	1.000	4.000	1.000	1
O9	1.000	1.000	1.000	0.897	3.897	0.960	2
O10	1.000	1.000	1.000	0.734	3.734	0.897	4
O11	1.000	1.000	0.745	1.000	3.745	0.901	3
O12	1.000	1.000	0.527	1.000	3.527	0.817	7
O13	1.000	0.601	1.000	1.000	3.601	0.845	6
a_{AI}	0.000	0.162	0.527	0.734	1.423	0.000	

Step 10. The [0,1]-scaled scoring system for negotiation offers is determined. First, the issue weights were calculated using the formula (7). We have the following: $w_{Concerts} = 1 - V(\text{O1})$ since O1 consists of the worst option for concerts. Therefore $w_{Concerts} = 1 - 0.612 = 0.388$. Consequently, $w_{Songs} = 1 - 0.675 = 0.325$, $w_{Royality} = 1 - 0.817 = 0.183$, $w_{Contract} = 1 - 0.897 = 0.103$. Next, the option ratings obtained in step 7 are rescaled using min-max normalization within each issue separately. For instance, the rating of option x_2^2=12 (songs) is determined as $v(x_2^2) = \frac{\tilde{L}_{O5} - \tilde{L}_{O4}}{\tilde{L}_{O8} - \tilde{L}_{O4}} = \frac{0.256 - 0.162}{1 - 0.162} = 0.112$. The results are presented in Table 5.

Table 5. Unfolding + MARS-based scoring system

Issue	Concerts				Songs					Royalties (%)				Contract		
Weights w_i	0.39				0.33					0.18				0.10		
Options	5	6	7	8	11	12	13	14	15	1.5	2	2.5	3	125	150	200
Scores $v\left(x_i^j\right)$	0.00	0.10	0.29	1.00	0.00	0.11	0.29	1.00	0.52	0.18	1.00	0.46	0.00	1.00	0.61	0.00

4 Conclusions and Future Research

In this paper, we proposed a novel approach for generating a scoring system for the negotiator based on the preference recommendations of experts. This approach mixes the notions of unfolding and MARS methods. MARS suggests the set of reference alternatives the experts should evaluate (rank order them). Unfolding visualizes the results of experts' recommendations and allows the evaluation of offers using a two-dimensional scale. It is a considerable advantage of the unfolding approach compared to other simple methods of deriving collective ranks (e.g., mean or dominant ranks). Further, it allows determining the cardinal ratings of offers using the MARS-based notions of the joint cardinal scale, which can then be decomposed, producing the negotiation issue weights and option scores. Consequently, the cardinal scoring system can be built and used later to support the negotiator in the actual and post-negotiation phases.

It is worth noting that the MARS requirements for building the H_{nRIS} set may result in quite an extensive list of reference offers in larger negotiation problems (i.e. when the number of issues increases). It may make the process of rank ordering them more difficult for experts and potentially result in various preference inconsistencies. A simple remedy to this problem is reducing the salient options within each issue. However, it will lead to a lower granularity of evaluation across the options, and consequently, some nuances in the shapes of marginal scoring functions may not be detected.

We showed an example of how our approach works using experimental data. In this example, the advantage of unfolding analysis could have also been observed. The unfolding graph allowed the negotiator to identify the outlier recommendations (e.g., the points identifying experts no. 23, 32, 33, and 51 in Fig. 2) that stick out of the remaining ones. The negotiator may wish to eliminate these outliers from the analysis, and then the unfolding analysis needs to be repeated, resulting in more precise evaluations as the W coefficient rises.

Further research in implementing this hybrid approach should focus on analyzing the impact of different dimensions of unfolding analysis on the final forms of scoring systems. It is clear that the higher dimensions allow unfolding analysis to reduce the representation errors of the dimensions calculated; however, simultaneously, it decreases the possibility of providing the informative visualization of these results for the negotiators [22].

Acknowledgements. This research was funded by grants from Bialystok University of Technology (WZ/WI-IIT/2/2022) and the University of Economics in Katowice.

References

1. Raiffa, H.: The Art and Science of Negotiation. Harvard University Press, Cambridge (1982)
2. Wachowicz, T., Roszkowska, E.: Holistic preferences and prenegotiation preparation. In: Kilgour, D.M., Eden, C. (eds.) Handbook of Group Decision and Negotiation, pp. 255–289. Springer, Cham (2021). https://doi.org/10.1007/978-3-030-49629-6_64
3. Keeney, R.L., Raiffa, H.: Decisions with Multiple Objectives: Preferences and Value Trade-Offs. Cambridge University Press, Cambridge (1993)
4. Mustajoki, J., Hämäläinen, R.P.: Web-HIPRE: global decision support by value tree and AHP analysis. INFOR: Inf. Syst. Oper. Res. **38**, 208–220 (2000). https://doi.org/10.1080/03155986.2000.11732409
5. Brzostowski, J., Roszkowska, E., Wachowicz, T.: Supporting negotiation by multi-criteria decision-making methods. Optim.-Studia Ekonomiczne. **5**, 59 (2012)
6. Kilgour, D.M., Chen, Y., Hipel, K.W.: Multiple criteria approaches to group decision and negotiation. In: Ehrgott, M., Figueira, J.R., and Greco, S. (eds.) Trends in Multiple Criteria Decision Analysis. pp. 317–338. Springer, Boston (2010). https://doi.org/10.1007/978-1-4419-5904-1_11
7. Kersten, G., Roszkowska, E., Wachowicz, T.: The heuristics and biases in using the negotiation support systems. In: Schoop, M., Kilgour, D.M. (eds.) GDN 2017. LNBIP, vol. 293, pp. 215–228. Springer, Cham (2017). https://doi.org/10.1007/978-3-319-63546-0_16
8. Kersten, G.E., Roszkowska, E., Wachowicz, T.: Representative decision-making and the propensity to use round and sharp numbers in preference specification. In: Chen, Ye., Kersten, G., Vetschera, R., Xu, H. (eds.) GDN 2018. LNBIP, vol. 315, pp. 43–55. Springer, Cham (2018). https://doi.org/10.1007/978-3-319-92874-6_4
9. Wachowicz, T.: Decision support in software supported negotiations. J. Bus. Econ. Manag. **11**, 576–597 (2010)
10. Górecka, D., Roszkowska, E., Wachowicz, T.: The MARS approach in the verbal and holistic evaluation of the negotiation template. Group Decis. Negot. **25**(6), 1097–1136 (2016). https://doi.org/10.1007/s10726-016-9475-9
11. Kersten, G., Roszkowska, E., Wachowicz, T.: An impact of negotiation profiles on the accuracy of negotiation offer scoring system? Exp. Study Multiple Criteria Decis. Making. **11**, 77–103 (2016)
12. Roszkowska, E., Kersten, G.E., Wachowicz, T.: The impact of negotiators' motivation on the use of decision support tools in preparation for negotiations. Int. Trans. Oper. Res. **33**, 1427–1452 (2023). https://doi.org/10.1111/itor.12995
13. Brzostowski, J., Roszkowska, E.: System rekomendacji doboru wag kryteriów oparty na ich charakterystyce probabilistycznej. Stud. Ekonomiczne **178**, 58–72 (2014)
14. Piasecki, K., Roszkowska, E., Wachowicz, T., Filipowicz-Chomko, M., Łyczkowska-Hanćkowiak, A.: Fuzzy representation of principal's preferences in inspire negotiation support system. Entropy **23**, 981 (2021)
15. Bennett, J.F., Hays, W.L.: Multidimensional unfolding: determining the dimensionality of ranked preference data. Psychometrika **25**, 27–43 (1960). https://doi.org/10.1007/BF02288932
16. Raiffa, H., Richardson, J., Metcalfe, D.: Negotiation Analysis: The Science and Art of Collaborative Decision Making. Harvard University Press, Cambridge (2002)

17. Kersten, G.E., Noronha, S.J.: WWW-based negotiation support: design, implementation, and use. Decis. Support Syst. **25**, 135–154 (1999)
18. Schoop, M.: Negoisst: Complex Digital Negotiation Support. Handbook of Group Decision and Negotiation, pp. 1149–1168 (2021)
19. Salo, A., Hämäläinen, R.P., Lahtinen, T.J.: Multicriteria methods for group decision processes: an overview. In: Kilgour, D.M., Eden, C. (eds.) Handbook of Group Decision and Negotiation, pp. 863–891. Springer, Cham (2021). https://doi.org/10.1007/978-3-030-49629-6_16
20. Flostrand, A.: Finding the future: crowdsourcing versus the delphi technique. Bus. Horiz. **60**, 229–236 (2017). https://doi.org/10.1016/j.bushor.2016.11.007
21. Moreno-Jiménez, J.M., Aguarón, J., Escobar, M.T., Salvador, M.: Group decision support using the analytic hierarchy process. In: Kilgour, D.M., Eden, C. (eds.) Handbook of Group Decision and Negotiation, pp. 947–975. Springer, Cham (2021). https://doi.org/10.1007/978-3-030-49629-6_51
22. Borg, I., Groenen, P.J.F., Mair, P.: Applied Multidimensional Scaling and Unfolding. Springer, Cham (2018). https://doi.org/10.1007/978-3-319-73471-2
23. Figuera, J., Greco, S., Ehrgott, M.: Multiple criteria decision analysis: state of the art. Springer, Boston (2016)
24. Jarke, M., Jelassi, M.T., Shakun, M.F.: MEDIATOR: towards a negotiation support system. Eur. J. Oper. Res. **31**, 314–334 (1987)
25. Wachowicz, T., Roszkowska, E.: Can holistic declaration of preferences improve a negotiation offer scoring system? Eur. J. Oper. Res. **299**, 1018–1032 (2022). https://doi.org/10.1016/j.ejor.2021.10.008
26. Larichev, O.I., Moshkovich, H.M.: ZAPROS-LM - A method and system for ordering multiattribute alternatives. Eur. J. Oper. Res. **82**, 503–521 (1995)
27. Bana e Costa, C., Vansnick, J.-C.: The MACBETH approach: basic ideas, software, and an application. In: Meskens, N., Roubens, M. (eds.) Advances in Decision Analysis, pp. 131–157. Springer, Dordrecht (1999)
28. Landis, J.R., Koch, G.G.: The measurement of observer agreement for categorical data. Biometrics 159–174 (1977)

A Group Decision-Aiding Protocol for Selecting a Post-industrial Cultural Tourism Product in Czeladź Commune in Poland

Tomasz Wachowicz(✉) and Marek Czekajski

University of Economics in Katowice, 1 Maja 50, 40-287 Katowice, Poland
tomasz.wachowicz@uekat.pl, marek.czekajski@edu.uekat.pl

Abstract. The study aims to propose a decision-aiding protocol for selecting the best cultural tourism product for a commune to promote its post-industrial heritage. It is a peculiar decision-making problem in which the alternatives have a composite character, i.e., they consist of many simple products that may present additional value combined. The decision-maker uses the opinion of many potential stakeholders regarding the alternatives evaluation, and these stakeholders may differ in their cognitive capabilities. Given the above, we propose an approach that produces a compromise recommendation regarding which product to choose under different preferences declared by the subsequent stakeholders. The problem is structured first, which involves defining a set of composite alternatives by an expert team and discussing them in a series of workshops with the stakeholders. They may also define additional evaluation criteria based on the predefined set of common goals. In the next step, the information processing styles of stakeholders are identified to suggest the multiple criteria decision aiding technique that best fits each stakeholder's cognitive capability. Finally, stakeholders' preferences are standardized to make them comparable on a common scale and aggregated using the adopted model of minimum cost consensus. We show how some elements of this approach work in a real-world problem of promoting the post-industrial heritage of the southern Polish commune Czeladź.

Keywords: Cultural tourism products · Preference elicitation · Cognitive capabilities · Group decision-making · Minimum cost consensus

1 Introduction

Cultural tourism product (CTP), as defined by World Tourism Organization, is "a combination of tangible and intangible elements, such as natural, cultural and man-made resources, attractions, facilities, services and activities around a specific center of interest which represents the core of the destination marketing mix and creates an overall visitor experience including emotional aspects for the potential customers. A tourism product is priced and sold through distribution channels and has a life-cycle" [1]. Cultural tourism and its products affect all areas of human economic activity. The development of the cultural tourism industry (e.g., entities, institutions, mutual relations, and products) plays a crucial role in creating a unique offer that attracts residents and tourists

Y. Maemura et al. (Eds.): GDN 2023, LNBIP 478, pp. 67–80, 2023.
https://doi.org/10.1007/978-3-031-33780-2_5

from outside a given commune, city, or region [2]. It also expands and diversifies the tourist offer of a given town or region thanks to the various forms of tourism products [3]. Most importantly, it allows to develop the territorial marketing and promote a given local government unit (LGU) [4, 5].

Contemporary product development based on a systems approach to cultural tourism [6] strongly emphasizes the advantages of systemic (complex) products or, in other words, multi-products. A systemic product combines various simple mono-products and their features, which may also have some synergic effect that better satisfies the consumers' needs and goals [7, 8]. Generally, there are seven main categories of tourism component mono-products that can be used in the process of creating such a systemic CTP, namely [9–13]: *things*, i.e., material goods or real products such as maps, guidebooks, souvenirs, promotional gadgets; *tourist services*, i.e., guided tour, accommodation, catering service; *sets of tourist services* (tourist rallies, cultural trip, etc.); *events* (festival, concert, cultural festivals); *architectural objects* (historic building, monument, architectural heritage); *routes* (walking trail /marked, unmarked/, bike route, questing trail); and *areas* (area with cultural and tourist heritage). Additionally, all CTPs may be designed in different forms of their physical instances such as real products (physical, material), digital (photos, videos, animations, graphics, e-books), multimedia (various digital resources as well as sound recordings, music, interactive elements), virtual (Internet portals or web-apps, computer programs and mobile applications), or mixed (that hybridize all the above).

The complexity and multidimensionality of the components of systemic tourism products make the process of their design and evaluation a non-trivial decision-making problem. Various evaluation criteria may be considered to decide how well they fulfill the goals of their owners. Additionally, potential decision makers (DMs) and stakeholders involved in CTP projects may trigger problems related to the conflict of views on the attractiveness of various CTP alternatives, which would require implementing tools for multiple criteria group decision-aiding (MCDA + GDM).

Several studies show formal techniques for analyzing cultural tourism planning and management problems. For example, Chou, Hsu, and Chen [14] applied a fuzzy multiple criteria decision making model to solve the decision problem regarding selecting a place for an international tourist hotel in Taiwan. In order to evaluate various variants of such a hotel location, 21 criteria have been created. The researchers presented the computational process and effectiveness of the model. Huang and Nguyen [15] also used the fuzzy approach to select optimal CTP promoting indigenous tribes' culture, in which the AHP and TOPSIS were combined. Wong and Fung [16] presented an interesting integration of the GIS system with the MCDA method. They showed that the GIS-based MCDA approach efficiently identifies potential sites for various ecotourism activities on Hong Kong's Lantau Island. Finally, Chang, Wey, and Tseng [17] researched the revitalization strategies selection problem for the historic Alishan Forest Railway in Taiwan. The integrated approach was based on fuzzy Delphi, analytic network process, and goal programming.

In this paper, we study a decision-making problem focused on designing and evaluating systemic CTPs when the opinions of various stakeholders need to be considered. More specifically, we design a protocol to identify CTPs out of the post-industrial remainings of Czeladź commune, a medium-sized town in the northern part of the Upper Silesian-Dąbrowa Basin Metropolis in Southern Poland. In the decision-making process, a DM is a collective body of the commune management board or the mayor's advising team. This DM seeks opinions regarding the attractiveness of various alternative CTPs that may be built out of the cultural heritage among all potential stakeholders. The stakeholders (employees of cultural institutions) differ in their cognitive capabilities and logical-mathematical and decision-making skills, making them more or less able to use one common MCDA tool efficiently and formulate a reliable opinion for DM. Therefore, the protocol includes a mechanism that identifies stakeholders' cognitive capabilities and suggests an adequate (cognitively adjusted) MCDA mechanism to ensure their preferences are elicited most reliably. Finally, the protocol provides solutions for standardizing stakeholders' preferences to make them comparable on a common scale and aggregating them, which is performed using the adopted model of minimum cost consensus. The collective preference information obtained this way is presented to the DM, who may use it in making the final decision or confront it with other recommendations, e.g., those provided by the potential consumers of such CTP. Our protocol may be generalized to other situations with CTPs evaluation in which similar constraints and assumptions regarding the stakeholders occur.

The paper is organized as follows. First, in Sect. 2, the post-industrial heritage of Czeladź is outlined, allowing us to identify the potential for building various CTPs and the stakeholders of the process. Next, in Sect. 3, the decision-making protocol is formalized as a step-by-step algorithm that identifies the problem, the stakeholders' cognitive capabilities, and the corresponding decision aiding methods and integrates them to produce the joint evaluation of identified CTPs. Finally, Sect. 4 shows how the algorithm works by specifying the CTPs out of the Czeladź post-industrial heritage (actual results from the onsite workshop) and eliciting preference information for some selected stakeholders (actual results and simulation).

2 Post-industrial Coal-Mining Heritage in Czeladź and the Stakeholders

Numerous traces, remains, monuments, non-historic architectural objects, and artifacts related to the post-industrial cultural heritage in the Czeladź Commune open the possibility of creating a thematic CTP. The post-industrial cultural heritage is related to the extensive, well-documented history of two former coal mines: "Saturn" and "Ernest-Michał" (later: "Czeladź" and "Milowice-Czeladź"), and the history of various infrastructure adjacent to these two coal mines. Obtaining substantive information regarding the post-industrial history of Czeladź is based primarily on the analysis of such sources as (1) the archival collection of the "Saturn" Museum in Czeladź [18, 19], (2) State Archive's resources regarding the Mining and Industrial Society "Saturn" [20], (3) State Archive's resources regarding the Nameless Society of Coal Mines "Czeladź" in Czeladź-Piaski

[18], (4) books and articles [21–25]. The identified post-industrial heritage base relates in general to:

1. Heritage of the former "Saturn" coal mine, namely:
   a. historic post-industrial facilities (mining shafts and machine rooms at the shafts etc.),
   b. historic machines and devices (steam power generator, compressor, control panel in the former mine power plant),
   c. interesting (in various aspects) other architectural objects (workshop buildings, assembly hall, boiler house),
   d. workers' patronage estates,
   e. houses for white-collar workers, skilled workers, officials,
   f. the building of the Board of the Mining and Industrial Society "Saturn",
   g. villa of the director of the "Saturn" mine,
   h. social buildings, school buildings, buildings for cultural purposes,
   i. recreational areas, parks, gardens,
   j. sports facilities (soccer fields),
   k. home gardens,
   l. information about non-existent objects, places, areas, etc. (e.g., powder magazine, sorting plant, railway siding).
2. Heritage of the former coal mine "Milowice-Czeladź", namely:
   a. workers' patronage estates,
   b. houses for white-collar workers, skilled workers, officials,
   c. houses of mining supervision employees,
   d. the chief mechanic's house,
   e. the villa of the mine director – Victor Viannay,
   f. mine management building,
   g. gardens, squares,
   h. clerical club,
   i. Neo-Romanesque parish church,
   j. school, and kindergarten buildings,
   k. information about industrial facilities, machines, and devices of the non-existent coal mine complex.

Considering the heritage above, the different categories of CTPs, and their different forms (real, virtual, etc.), one faces the problem of designing alternatives of complex CTPs, which are composed of different types of CTP and different types of post-industrial heritage. It involves many issues to be resolved regarding important functions and aims of the planned complex CTP; potential mixes of categories, types, forms, and instances of such a new CTP; its important features and attributes, etc. The examples of predefined alternatives of such complex CTP that were designed using the support protocol we proposed are shown later in Table 2 (Sect. 4).

In creating a local/regional CTP, various entities play an important role – both from the environment of the local government, as well as entities of culture, tourism, educational institutions, or non-governmental, non-profit organizations. Based on the analysis of the scope of competencies of employees of various entities operating in the Czeladź

Commune, five categories of stakeholders who will participate in creating a CTP can be identified (Table 1).

Table 1. Stakeholders in the process of creating a new complex CTP in Czeladź

Category of stakeholders	Names of the stakeholders
I. Stakeholders at the level of the LGU of the Czeladź Commune	1) Formal decision-makers: Mayor of Czeladź and his deputies 2) Substantively competent employees of organizational units (offices, departments) of the Municipal Office in Czeladź, whose scope of activities concerns the promotion of LGU, territorial marketing, culture, tourism, city development, etc
II. Stakeholders in units, entities and institutions subordinate to the Czeladź Commune	1) Formally decision-makers in municipal cultural institutions – directors, managers, etc 2) Formal decision-makers in municipal institutions related to tourism, sport, and recreation – directors, managers, etc 3) Formally decision-makers in other entities whose statutory activity is related to culture, tourism in general and cultural tourism
III. Stakeholders in non-governmental organizations	Substantively competent employees of non-governmental organizations whose statutory activity is related to the promotion of culture, tourism, and cultural tourism – decision-makers in local non-governmental organizations in the Czeladź Commune
IV. Stakeholders concentrated on other forms of social and public activity	Local citizen initiatives, informal action groups, etc
V. Stakeholders in municipal educational institutions (related to the subject of cultural tourism)	1) Directors and/or their alternates 2) History, cultural studies, social studies, basics of marketing, basics of entrepreneurship teachers etc

3 Decision-Aiding Protocol for Selection of Best Post-industrial CTP

The heritage outlined in Sect. 2 and the decision-making context described in Sect. 1 expose a multi-criteria and multi-stakeholder character of the problem under consideration. Therefore, we decided to design the group decision-aiding protocol that could be used to support DMs in the Czeladź Commune in investigating the preferences of potential stakeholders regarding the problem of building a new CTP promoting post-industrial heritage. It was designed based on preliminary theoretical studies and on-the-spot consultations with the experts from the "Saturn" Museum in Czeladź (considered one of the key stakeholders and potential coordinators of a future post-industrial CTP project). The protocol consists of the following six steps:

1. *Pre-structuration of the CTP problem.* A group of experts (X) set up as a principal coordinator of the CTP project follows the first steps of the PrOACT [26] approach to structure the decision-making problem. The group is supported by the analytic unit (a decision-making methodologist and a data analyst), which may use some interactive facilitation techniques such as mind or cognitive mapping [27]). As a result, a list of alternatives is defined $A = \{A_1, \ldots, A_m\}$ that consists of a precise description of each proposed CTP and its components, as well as the list of predefined evaluation criteria $G = \{G_1, \ldots, G_n\}$ with a detailed vocabulary. Thus, a CTP problem is defined as $P = \{A, G\}$.
2. *CTP problem consultations with stakeholders.* The predefined problem P is disseminated to all stakeholders (individuals or groups). Then the analytic unit supports a series of workshops, one for each stakeholder S_k ($k = 1, \ldots, K$), during which the suggestions for new alternatives may be formulated (A^{S_k}). Similarly, individual evaluation criteria may be raised, or some predefined ones may be considered irrelevant, which will lead to defining the stakeholder-specific sets of criteria G^{S_k}.
3. *Restructuring CTP problem.* The X-team gathers the stakeholders' recommendations A^{S_k} and G^{S_k}. The new joint set of alternatives A^* is built that takes into consideration the original set A and the stakeholders' recommendations, provided some similar individual recommendations are first tuned up to formulate a common alternative, i.e.

$$A^* = \{A_i^*\} = A \cup \left\{ \bigcup_{k=1}^{K} A^{S_k} \right\}. \tag{1}$$

A new restructured CTP problem may be then formulated as $P^R = \left\{A^*, \{G^{S_k}\}_{k=1,\ldots,K}\right\}$.

4. *Identifying the stakeholders' cognitive profiles.* Cognitive style is a subjective process by which people perceive, organize and use information during decision-making [28]. It was proven to affect the DMs' efficiency in using the decision-aiding tools for preference elicitation (see, e.g., [29]). Therefore we identify the stakeholders' cognitive (decision-making) profiles CP^{S_k} through a preselected psychometric inventory and try to assign to each profile the cognitively most fitting decision-aiding technique. Following the earlier studies of Roszkowska and Wachowicz [30], we suggest using REI-20 inventory [31] since the resulting profile may easily be linked to three clusters of profiles for which three straightforward MCDA techniques – best evaluated by the DMs and resulting in adequate preference representations – were identified. Generally, the SMART technique is suggested for DMs with the dominating rational mode, while for those with balanced rational and experiential modes or more intuitive ones, AHP seems quite effective. Finally, the avoidant DMs may use TOPSIS with a corresponding pictogram-based interface.

5. *Preference analysis of individual stakeholders.* With their profiles defined, the stakeholders enter the preference elicitation phase supported by the technique most suited to their profile. As a result, the scoring systems SS^{S_k} are defined for the stakeholders, which specify the cardinal performance of all offers under consideration (from A^*), i.e.

$$SS^{S_k} = \left\{ v^{S_k}(A_i^*) \right\}_{i=1,\ldots,|A^*|}. \tag{2}$$

As SS^{S_k} may be determined from different MCDA methods, the mechanism needs to be proposed to ensure they will finally operate with the same type of scale to make further comparisons of scores across the stakeholders possible. Since TOPSIS and SMART may be scaled to [0; 100]-range of interval interpretation easily, the only problem occurs when the AHP ratio-scale results must be integrated. Thus, we suggest using an AHP extension proposed for scoring large problems [32], in which a simple rescaling of single-criterion performances is suggested.

When the voices of potential customers must also be considered, a web-based mechanism for gathering their preferences might be designed. A complete preference elicitation approach might be used for each of them (as for regular stakeholders), or – given rather a significant number of them – a more straightforward method might be implemented, in which each defines a simple order of some selected subset of A^*. Then, using a mixed approach implementing the MARS technique and unfolding analyses (as suggested by [33]), the collective scoring system of customers SS^C may be produced.

6. *Aggregating stakeholders' preference information.* Previously performed steps allow defining the CTP problem as the following n-tuple

$$CTP = \left\{ A^*, \left\{ G^{S_k}, CP^{S_k}, SS^{S_k} \right\}_{k=1,\ldots,K}, SS^C \right\}. \tag{3}$$

The analytic team now generates a collective evaluation of potential CTP alternatives for DM that aggregates various opinions of all the stakeholders. Assuming that the DM may consider that the importance of various stakeholders' recommendations may differ, we suggest implementing the weighted minimal cost consensus model derived from Ben-Arieh's initial approach [35], although other notions of consensus may be applied. Here we will search for a compromise solution that would minimize the following cost function:

$$\min f(v') = \sum_{k=1}^{K+1} \sum_{i=1}^{|A^*|} w_k c_k \left| v^{S_k}(A_i^*) - v'(A_i^*) \right|, \tag{4}$$

where $v'(A_i^*)$ is a consensus score of offer A_i^*, w_k and c_k are weight and consensus cost of k th stakeholder, respectively, and index $k = K + 1$ denotes the collective stakeholder with the preferences described by the scoring system SS^C.

4 CTP Decision-Aiding Protocol in Designing and Evaluating Czeladź Post-industrial CTP

Creating a new post-industrial CTP in Czeladź is still an ongoing project. As for now, we only have partial data gathered from interacting with one selected stakeholder, the representative of the promotion unit of "Saturn" Museum (stakeholder 1). In this work, we have also considered further two stakeholders, critically important for the project, i.e., the head of the City Promotion, Culture, and International Cooperation Department of the Czeladź City Hall (stakeholder 2) and the manager of the "Mine of Culture" – a cultural institution from Czeladź (stakeholder 3). Their preferences were obtained using a role-playing-like approach, in which stakeholder 1 defined his presumptions regarding the interests of stakeholders 2 and 3. Below we show the CTP design and evaluation process performed according to the protocol outlined in Sect. 3.

Step 1. We organized a workshop with stakeholder 1, during which a problem was defined and the set of alternatives specified. In defining the problem and alternatives, we used the detailed checklist of questions recommended by the standard PrOACT procedure [26]. Consequently, set A consisting of 10 predefined alternatives defining complex CTPs promoting post-industrial heritage were defined, which is shown in Table 2.

Next, the decision criteria were defined. Again, many perspectives exist on how these criteria may be formulated [36–38]. Following these recommendations, the expert team prepared a list of predefined, unified criteria (set G):

- G_1: Attractiveness of the product from the point of view of tourists.
- G_2: Innovation in product development.
- G_3: New technologies used in product development and its promotion.
- G_4: Economic and social importance for the development of the region.
- G_5: Relationships to events or traditions related to post-industrial times.
- G_6: Authenticity (how well the product describes the post-industrial time).
- G_7: Uniqueness (how original the product is).
- G_8: Impact on general tourist infrastructure of the region.
- G_9: Stimulation of tourist events in the region.
- G_{10}: Stimulation of cultural events.
- G_{11}: Providing new experiences, emotions, and social contacts.
- G_{12}: Enhancing promotion of the region, creating the region's image.
- G_{13}: Providing educational impact for the users.
- G_{14}: Shaping local/regional identity.

Steps 2 and 3. Stakeholders had to revise the problem and decide on evaluation criteria that best fit the specifics and context of the CTP creation process related to promoting post-industrial heritage in Czeladź. Without losing the generality of our approach, we will assume that all three stakeholders decided to evaluate the original set of alternatives proposed by the expert team and use a common subset of the criteria form G. As a result, we obtained a restructured problem defined in the following form:

$$P^R = \left\{A^* = A, G^{S_1} = G^{S_2} = G^{S_2} = \{G_3, G_4, G_8, G_{10}, G_{11}, G_{12}, G_{13}\}\right\}. \quad (5)$$

Table 2. Predefined alternatives for Czeladź post-industrial CTP project.

Alternative	Type of CTP in relation to the multidimensionality of the product	Description of the alternative of CTP
1. Route of "Postindustria"	Product-route in real form and/or Product-thing in a hybrid form	Thematic cultural route leading through the most important points (places) of post-industrial heritage. The route also consists of dedicated, thematic sub-routes and educational trails concerning the technical monuments (machines, devices) and residential architecture
2. "Postindustria" Family Festivities	Product-event in a real form and/or Product-event in a hybrid form	Thematic tourist and cultural events containing educational workshops, outdoor family games, do-it-yourself (DIY) workshops, and multimedia presentations of places, traces, and artifacts
3. "Postindustria" Family Rally	Product-services set in real form	Thematic annual sports, tourist, and culture event with elements of learning (workshops) about post-industrial culture
4. "Postindustria" Quest of Czeladź	Product-route in real form with questing and/or product-route in hybrid form with questing	Questing for post-industrial cultural heritage; outdoor game solving puzzles, tasks, quizzes, and finding the password
5. "Postindustria" Museum	Product-object in real form and/or product-object in hybrid form	Temporary, cyclical (once a year) exhibitions at the "Saturn" Museum and Contemporary Art Gallery "Elektrownia"
6. "Terra Postindustria"	Product-area in the real form	Thematic geographically determined area of the former two coal mines, their patron estates, and other infrastructure sites with routes, trails, questing games, and cultural tourism facilities

(*continued*)

Table 2. (*continued*)

Alternative	Type of CTP in relation to the multidimensionality of the product	Description of the alternative of CTP
7. "Postindustria Story"	Product-service in real form and/or Product-service in hybrid form	Thematic story-based guided tour of the entire area related to the two mines and their heritage, divided into several thematic sections: (1) technical monuments, (2) residential architecture, (3) recreation, entertainment, and (4) everyday life of mine workers, customs, rituals
8. Portfolio product A	Material good (thing) + service + route	Map of post-industrial attractions, guided service of the most important attractions, and a thematic route through the most important post-industrial attractions
9. Portfolio product B	Event + services set + virtual route	Thematic tourist and cultural festivities, picnics, festivals, exhibitions, etc. Thematic, sports, and tourist rallies with elements of learning about the post-industrial culture. Virtual route on the web
10. Portfolio product C	Product-thing in multimedia form + virtual service + virtual route	Interactive map of attractions (with photos, videos, graphics, animations), including a virtual tour combined with the audiobooks through the virtual route on the website

Step 4. Stakeholder 1 was asked to complete the REI-20 test (in Polish) to determine his cognitive style. His average rationality and experientiality modes were 2.8 and 3.1, respectively. According to Roszkowska and Wachowicz [30], he can be classified as versatile (or slightly experiential) DM, for which they suggest applying the AHP technique for eliciting preferences. To show how the group aggregation works in the later steps, we will assume that stakeholder 2 was classified as avoidant, while stakeholder 3 was rational (analytical). Consequently, TOPSIS was used for the former, while SMART was used for the latter.

Step 5. For stakeholder 1, AHP was used to analyze his preferences. Since the number of alternatives was too big to implement the classic pair-wise comparisons without risking a fatigue effect that could affect the consistency of preference declarations, we implemented the AHP-express approach [39] to find the single criteria priorities across the alternatives. Next, the AHP results were normalized, as suggested by the procedure for large-scale AHP problems. It required that the scores describing the single-criterion performances of alternatives derived from the eigenvectors be normalized using the max-min scaling procedure [32]. It ensured that the evaluation scales and results obtained by all three stakeholders could be compared. An example of AHP analysis and normalization is shown in Table 3.

Table 3. AHP-express analysis and normalized scores of alternatives for criterion G_3

	A_1	A_2	A_3	A_4	A_5	A_6	A_7	A_8	A_9	A_{10}	Sum
A_1	1	3	2	2	2	1	3	5	2	2	
1/a	1.00	0.33	0.50	0.50	0.50	1.00	0.33	0.20	0.50	0.50	5.36
Priorities*)	0.19	0.06	0.09	0.09	0.09	0.19	0.06	0.04	0.09	0.09	
Normalized priorities	1.00	0.17	0.38	0.38	0.38	1.00	0.17	0.00	0.38	0.38	

*) priorities are determined from AHP-express formula$pr_j = \frac{1}{a_{ij}\sum_k 1/a_{ik}}$

As a result, the global ratings $v^{S_1}(A_i^*)$ were [0; 1]-scaled. For the remaining stakeholders, we used classic TOPSIS and SMART analyses. Since the results of the latter one are [0; 100]-scaled, we divided $v^{S_3}(A_i^*)$ by 100 to ensure cross-stakeholders' comparability at the interval-scale level. The global evaluations of offers from A^* for each stakeholder are shown in Table 4 (columns 2 – 4).

Table 4. Assessments of CTP alternatives promoting post-industrial heritage in Czeladź

Alternative	v^{S_1}	v^{S_2}	v^{S_3}	Consensus score
1) Route of "Postindustria"	0.941	0.057	0.735	0.635
2) "Postindustria" Family Festivals	0.886	0.513	0.385	0.674
3) "Postindustria" Family Rally	0,798	0.570	0.350	0.640
4) "Postindustria" Quest of Czeladź	0.621	0.599	0.545	0.599
5) "Postindustria" Museum	0.450	0.298	0.225	0.359
6) "Terra Postindustria"	0.358	0.624	0.385	0.443
7) "Postindustria Story"	0.227	0.327	0.130	0.238
8) Portfolio product A	0.135	0.670	0.580	0.385
9) Portfolio product B	0.136	0.707	0.730	0.426
10) Portfolio product C	0.010	0.684	0.755	0.361

Step 6. Having the global evaluations of CTPs for all stakeholders determined, we implemented the notion of minimum cost consensus. We assume that the DM considers that the gravities of recommendations provided by stakeholders differ. He assigns the following priorities to the stakeholders: 0.5, 0.3, and 0.2. Using these priorities as conceding costs, we build the consensus cost function out of v^{S_1}, v^{S_2} and v^{S_3} values using formula (4). Minimizing it allows us to determine the minimum cost consensus and corresponding global priorities of offers (shown in column 5 in Table 4). The best solution is identified this way, which is A_2 – "Postindustria" Family Festivals. It consists of product-event, product-services set, and product-thing. An example of such a complex product can be thematic tourist-cultural events with educational workshops, outdoor family games, DIY workshops, multimedia presentations of places, traces, and artifacts, as well as a concert of music in a post-industrial mood. However, one should also note that two other alternatives, A_1 (route) and A_3 (rally), have a very similar consensus score. DM should ask for the sensitivity analysis showing how they may change in the rank order while some minor changes in stakeholders' preference declarations are assumed. He may also ask other stakeholders to cast their opinions regarding CTPs under consideration or use the crowdsourcing method to consider the consumers' collective voice.

5 Conclusions

In this paper, we have proposed a comprehensive design of group decision-aiding protocol that could be used for choosing the systemic CTP out of the consensual recommendations provided by many experts. The multitude of experts is typical to designing and implementing CTP by any local authority, which may wish (or be obliged by law) to consult the idea of such CTP with potential stakeholders and learn their opinion about the possible solutions. Therefore, we proposed introducing a pre-decision-making structuration phase performed by the expert group to make further problem definition by the stakeholders easier. Since the evaluation of CTP is usually multi-criteria, the stakeholders may not have adequate formal or decision-analytic skills to formulate their recommendations reliably and require decision-aiding support. Therefore, we decided to recognize stakeholders' cognitive abilities and elicit their preferences using the cognitively best-fit MCDA method. Simplifying AHP-based preference analysis and standardizing the results obtained from different methods were also suggested to ensure the scores were provided on a common and comparable scale. A simple aggregation of stakeholders' views based on the minimum cost consensus model was finally proposed.

We showed how the protocol might work, analyzing the problem of building post-industrial CTP for Czeladź Commune. We used actual data from one stakeholder and simulated some recommendations for two others to show how the results of various types may be easily aggregated into a collective recommendation for a DM. Further research will focus on organizing decision-aiding workshops according to the protocol proposed with other stakeholders and providing a final group recommendation of post-industrial CTP to the mayor of Czleadź commune. We will also try to develop a comprehensive procedure for sensitivity analysis. It will consider the potential changes in recommendations depending on various trades within the individual scoring systems of subsequent stakeholders as well as the changes in the parameters of the cost consensus function (including the threshold for no-cost concessions).

References

1. Tourism and Culture I UNWTO. https://www.unwto.org/tourism-and-culture
2. Kay Smith, M., Pinke-Sziva, I., Berezvai, Z., Buczkowska-Gołąbek, K.: The changing nature of the cultural tourist: motivations, profiles and experiences of cultural tourists in Budapest. J. Tour. Cult. Chang. **20**, 1–19 (2022)
3. Bec, A., Moyle, B., Schaffer, V., Timms, K.: Virtual reality and mixed reality for second chance tourism. Tour. Manage. **83**, 104256 (2021)
4. Panasiuk, A.: Policy of sustainable development of urban tourism. Polish J. Sport Tourism **27**, 33–37 (2020)
5. Felsenstein, D., Fleischer, A.: Local festivals and tourism promotion: the role of public assistance and visitor expenditure. J. Travel Res. **41**, 385–392 (2003)
6. Jakulin, J.T.: Systems approach to cultural tourism and events. Acad. Turistica - Tourism Innov. J. **12**, 185–191 (2019)
7. Żabiński, L.: Marketing of systemic products. PWE, Warszawa (2012) (in Polish)
8. Kotler, P.: Principles of Marketing. Financial Times Prentice Hall, Hoboken (2005)
9. Yu, X., Xu, H.: Cultural heritage elements in tourism: a tier structure from a tripartite analytical framework. J. Destin. Mark. Manag. **13**, 39–50 (2019)
10. Burkart, A., Medlik, S.: Tourism: Present, Past, and Future (1974)
11. Mason, P.: Tourism Impacts, Planning and Management. Routledge, Milton Park (2020)
12. Medlik, S., Middleton, V.: Product formulation in tourism. Tourism Mark. **13**, 173–201 (1973)
13. Stokes, R.: Tourism strategy making: insights to the events tourism domain. Tour. Manage. **29**, 252–262 (2008)
14. Chou, T.-Y., Hsu, C.-L., Chen, M.-C.: A fuzzy multi-criteria decision model for international tourist hotels location selection. Int. J. Hosp. Manag. **27**, 293–301 (2008)
15. Huang, F.-H., Nguyen, H.: Selecting optimal cultural tourism for indigenous tribes by fuzzy MCDM. Mathematics **10**, 3121 (2022). https://doi.org/10.3390/math10173121
16. Wong, F.K., Fung, T.: Ecotourism planning in Lantau Island using multiple criteria decision analysis with geographic information system. Environ. Plann. B. Plann. Des. **43**, 640–662 (2016)
17. Chang, Y.-H., Wey, W.-M., Tseng, H.-Y.: Using ANP priorities with goal programming for revitalization strategies in historic transport: a case study of the alishan forest railway. Exp. Syst. Appl. **36**, 8682–8690 (2009)
18. The Nameless Society of the Coal Mines "Czeladź" in Czeladź-Piaski (1919) (in Polish)
19. Archival collection of the "Saturn" Museum in Czeladź. Project of the mine building (in Polish)
20. The Mining and Industry Society "Saturn" (1862) (in Polish)
21. Lazar, S., Binek-Zajda, A.: Patronage estate "Piaski". History and architecture. Czeladź (2015) (in Polish)
22. Binek-Zajda, A., Lazar, S., Szaleniec, I.: "Saturn" mine and workers' housing estate. History, architecture, people. Towarzystwo Powszechne „Czeladź" and Muzeum Saturn w Czeladzi, Czeladź (2016) (in Polish)
23. Chmielewska, M., Lamparska, M., Pytel, S., Jurek, K.: Patronage estates of Zagłębie. Tourist trail design. Association for the Protection of Natural and Cultural Heritage "MOJE MIIASTO", Będzin (2016) (in Polish)
24. Domaszewski, K.: From a trip to Saturn. "Zeszyty Czeladzkie" (2000) (in Polish)

25. Kurek, R.: Beginnings and development of industry in Czeladź. In: Drabina, J. (ed.) History of Czeladź, Czeladź (2012) (in Polish)
26. Hammond, J.S., Keeney, R.L., Raiffa, H.: Smart Choices: A Practical Guide to Making Better Decisions. Harvard Business Review Press, New York (2015)
27. Eden, C., Ackermann, F.: Cognitive mapping expert views for policy analysis in the public sector. Eur. J. Oper. Res. **152**, 615–630 (2004). https://doi.org/10.1016/S0377-2217(03)00061-4
28. Kozhevnikov, M.: Cognitive styles in the context of modern psychology: toward an integrated framework of cognitive style. Psychol. Bull. **133**, 464–481 (2007). https://doi.org/10.1037/0033-2909.133.3.464
29. Roszkowska, E., Kersten, G.E., Wachowicz, T.: The impact of negotiators' motivation on the use of decision support tools in preparation for negotiations. Int. Trans. Oper. Res. **30**, 1427–1452 (2023). https://doi.org/10.1111/itor.12995
30. Roszkowska, E., Wachowicz, T.: Cognitive style and the expectations towards the preference representation in decision support systems. In: Morais, D.C., Carreras, A., de Almeida, A.T., Vetschera, R. (eds.) GDN 2019. LNBIP, vol. 351, pp. 163–177. Springer, Cham (2019). https://doi.org/10.1007/978-3-030-21711-2_13
31. Pacini, R., Epstein, S.: The relation of rational and experiential information processing styles to personality, basic beliefs, and the ratio-bias phenomenon. J. Pers. Soc. Psychol. **76**, 972–987 (1999). https://doi.org/10.1037/0022-3514.76.6.972
32. Forman, E.H., Selly, M.A.: Decision by Objectives. World Scientific Publishing, Singapore (2001)
33. Wachowicz, T., Roszkowska, E., Filipowicz-Chomko, M.: Using unfolding analysis and MARS approach for generating a scoring system from a group preference information. In: LNBiP, Proceedings of the GDN Conference (2023)
34. Borg, I., Groenen, P.J.F., Mair, P.: Applied Multidimensional Scaling and Unfolding. Springer, Cham (2018). https://doi.org/10.1007/978-3-319-73471-2
35. Ben-Arieh, D., Easton, T.: Multi-criteria group consensus under linear cost opinion elasticity. Decis. Support Syst. **43**, 713–721 (2007). https://doi.org/10.1016/j.dss.2006.11.009
36. Ramírez Guerrero, G., García Onetti, J., Chica Ruiz, J.A., Arcila Garrido, M.: Concrete as heritage: The social perception from heritage criteria perspective: The eduardo Torroja's work (2020)
37. Abdurahman, A.Z.A., Ali, J.K., Khedif, L.Y.B., Bohari, Z., Ahmad, J.A., Kibat, S.A.: Eco-tourism product attributes and tourist attractions: UiTM undergraduate studies. Procedia Soc. Behav. Sci. **224**, 360–367 (2016)
38. Fuadillah, N., Murwatiningsih, M.: The effect of place branding, promotion and tourism product attribute on decision to visit through the destination image. Manage. Anal. J. **7**, 328–339 (2018)
39. Leal, J.E.: AHP-express: A simplified version of the analytical hierarchy process method. MethodsX. **7**, 100748 (2020). https://doi.org/10.1016/j.mex.2019.11.021

Capturing Stakeholder Value Drivers in Participatory Decision Analysis

Andreas Paulsson[1(✉)] and Aron Larsson[1,2]

[1] Department of Computer and Systems Sciences, Stockholm University, SE-16440 Kista, Sweden
{apaulsson,aron}@dsv.su.se

[2] Department of Information Systems and Technology, Mid Sweden University, SE-85170 Sundsvall, Sweden

Abstract. A fundamental element of participatory group decision processes is acknowledging the desires and concerns of the participating stakeholders. That involves reaching out to stakeholders and asking them to provide input, whereby their desires and concerns can be addressed in the decision process. In this paper, we elaborate on a case of a participatory decision process intending to form a municipal growth strategy in northern Sweden, where an increased understanding of stakeholder values was at the forefront of the project. We present the concept of *value driver* as a means for interpreting and structuring stakeholder value input in participatory processes for decision-makers to gain an increased understanding of the stakeholders' desires and concerns. In particular, we discuss the aggregation of such value drivers when reaching out to a large set of stakeholders via surveys and how such an approach can inform a participatory decision analysis process. The aim is to provide a conceptual representation of stakeholder values that can inform participatory decision processes seeking compromise solutions, such as how municipal resources should be allocated effectively based on what residents and business representatives find important for living, working, and running businesses in a municipality.

Keywords: Group decision · Participation · Stakeholder value · Value driver

1 Introduction

One of the more fundamental aspects of any group decision analysis process is the formulation of shared objectives, i.e., understanding what creates value for the stakeholders. In some cases, a decision is prepared and analyzed on behalf of others, for instance, in municipal planning or other public domains. In such cases, a participatory approach to decision-making is considered valid. The decision maker obtains the basis for formulating the objectives by reaching out to stakeholders, asking for their inputs and perspectives, see, e.g., [11]. One rationale for participatory decision processes is the assumption that by allowing stakeholders

Y. Maemura et al. (Eds.): GDN 2023, LNBIP 478, pp. 81–93, 2023.
https://doi.org/10.1007/978-3-031-33780-2_6

to provide their views, a more comprehensive understanding of the issues they face can be obtained by a decision-making body, thereby facilitating an increase in decision quality [1]. This paper reports on further elaboration of participatory decision practices in the Swedish municipality of Kramfors. The municipality undertook participatory methods for creating a new municipal growth strategy, focusing on including the views and desires of, in particular, youths, businesses, and households to improve the capability of the municipal decision-makers to simulate positive influence on its citizens and businesses.

One approach to collecting and analyzing stakeholder values involves interviewing stakeholder representatives and designing a value tree with all stakeholder groups' values represented [4]—performance attributes represent the stakeholders' values which are not defined otherwise. Weights, expressing the relative importance of attributes, are subsequently assigned to the attributes with respect to some predefined set of decision alternatives, cf. [3]. However, suppose we are to follow the idea behind value-focused thinking [8], as the authors of this report purport to do. In that case, the stakeholders' values must be collected and analyzed before any alternatives are present. As a result, those values cannot represent the performance attributes of the decision alternatives as we have yet to develop the latter—for the same reason, we cannot elicit attribute weights.

In this study of value drivers, we distinguish between *values of the public* and *public values*. The values of the public are the things that the citizens explicitly value, according to themselves, in a given context. Public values, on the other hand, are the values generally considered important for society by some ruling paradigm of authority or culture and adopted by policymakers. Public values are typically fundamental and functional societal values, such as having access to clean drinking water, energy, security, etc. Public values are essentially a product of the values held by the relevant policymakers, based on beliefs about what serves society the best. However, for someone who values staying in power, the definition of public values may be different for someone altruistically driven and interested in some natural and biological well-being of the citizens. Values collected from the citizens within the context of a public decision process, on the other hand, might not include fundamental societal value drivers such as access to clean drinking water and fire departments but instead values associated with a particular decision process. Thus, an important task of community leadership is taking the elementary and essential public values, whose importance would become severely apparent to the public should they be absent, into account, along with the values of the public collected from citizens in participatory decision processes.

Realistically, there should always be an apparent reason for retrieving and interpreting the values of the public. That reason will play an important role in subsequent analyses of collected data, e.g., how the collected values will be interpreted and structured. A relatively generic reason for taking the values of the public into account when making strategic decisions, e.g., in a public body such as a municipality, is that the chance of a thriving society increases as decisions align with the values of the citizens and presumable citizens. A poor alignment will, over time, most likely result in people losing trust in participatory practices.

A large number of stakeholders requires both a structured and a non-exhausting approach to data collection to obtain a representative number of stakeholder views. This paper aims to enhance the process of collecting, interpreting, and structuring the values of the public by introducing a concept we refer to as value drivers to the benefit of participatory decision-making processes.

2 Value, Valence, and Experience

In this paper, we consider value to be something that is discovered through the valence of experiences of an individual. In a decision situation, having a set of alternatives available before us, our choice will be determined by the valence-signals "generated by our evaluative systems responding to representations of those alternatives" [2]. According to this account, valence signals are affected not only by imagined future experiences, i.e., the consequences assumed to follow a particular choice, but also by the uncertainty accompanying each such possible consequence.

We assume the valence of an experience to be a non-conceptual representation of value, as in [2]. As such, we see value as something analogous to a governor—which can adapt over time—of the reactive system (i.e., the amygdala [9]) that produces valence signals in response to experiences, imagined or real.

Let $E = E_I \cup \{e_c\}$ be the set of experiences currently available to an individual. Here, E_I is the set of imagined experiences and e_c is the current experience. We then stipulate that for any experiences $e_i, e_j, e_k \in E$, the value of e_i is greater than the value of e_j if and only if the valence toward e_i is more positive than the valence toward e_j, and the value of e_i is equal to the value of e_j if and only if the valence toward e_i and e_j is the same. Similarly, the difference in value between e_i and e_j is greater than the difference in value between e_j and e_k if and only if the difference in valence toward e_i and e_j is more pronounced than the difference in valence toward e_j and e_k.

The value of alternatives involving uncertainty is assumed to be similarly related to valence, even though such alternatives are outside the scope of this study. As a consequence, the value ordering of a set of alternatives need not coincide with the preference ordering deemed rational by choice rules, such as maximizing the expected value.

We assume an experience to be the consciousness of something [6], more specifically, to the *pure essence* of things as essential beings, whether actual, historical, or given by imagination. Take an experience of a situation s which involves watching a game in an ice hockey arena. In reality, s involves several minutiae that would be impossible for anyone to be aware of, such as the breathing of almost everyone else in the arena. Consequently, one can only be conscious of parts of s, not the whole. Also, one cannot be aware of the backside of the puck (except indirectly, through a mirror or a display). The same principle applies to every element of s. In any case, something we experience about s makes it a situation in which we watch a game of ice hockey—that is the essence of s. To some extent, we could modify s, yet the essence of an ice hockey game would remain. Removing some spectators, for example, would typically not change an ice hockey game into something else. The *essential truths* of an experience given to us bring about the part of the pure essence of that experience that is accessible to our consciousness.

Let $E_{i,t}$ be the set of experiences accessible to individual a_i at time t. Furthermore, each experience $e \in E_{i,t}$ satisfies one or more essential truths $\mathcal{P}_e = \{P_j\}$, which in turn defines the pure essence of e. The pure essence of an experience is obtained through essential intuition or ideation [6], at varying degrees of universality depending on the extent to which the pure essence indeed can be captured. For each pure essence $\mathcal{P}$, whether partial or complete, we can attach possible instances of factual situations or thought-up experiences that have $\mathcal{P}$ as a subset of its essence. Note that some experiences may lack corresponding actual instances, such as a fantasy where one can breathe underwater without the help of any technical gear in the world as we now know it.

The more substantial the pure essence of an experience, i.e., the more essential truths an experience satisfies, the more specialized or concrete that experience becomes. If $\mathcal{P} \subset \mathcal{P}'$ and any essential truth in $\mathcal{P}' \setminus \mathcal{P}$ has a smaller extension than any essential truth in $\mathcal{P}$, then $\mathcal{P}'$ is a specialization of $\mathcal{P}$. The greater the extension of an essential truth, the greater its degree of universality. Hence, the degree of universality of an experience e depends on the degree of universality of its essential truths. Let $u(\mathcal{P}_e) = \bigcap E_{P_k}$ for all $P_k \in \mathcal{P}_e$, where E_{P_k} is the extension of P_k, be the universe of $\mathcal{P}_e$. Then let $d(\mathcal{P}_e) = |u(\mathcal{P}_e)|$ be the degree of universality of $\mathcal{P}_e$.

At some level of universality, two or more individuals can share similar experiences. The more the individuals have in common (e.g., in terms of past experiences, education, and culture), the more detailed essential truths their experiences would have in common due to the stock of knowledge at hand [12], i.e., the more specific the level of common universality they would naturally be able to obtain. If a pair of individuals experience e and e' respectively from a situation s, with one of the individuals obtaining the pure essence $\mathcal{P}_e$, and the other $\mathcal{P}_{e'}$, then they are *essence consistent* to degree $(\mathcal{P}_e \cap \mathcal{P}_{e'}) \,/\, (\mathcal{P}_e \cup \mathcal{P}_{e'})$, and *universality consistent* to degree $(u(\mathcal{P}_e) \cap u(\mathcal{P}_{e'})) \,/\, (u(\mathcal{P}_e) \cup u(\mathcal{P}_{e'}))$.

Since the precise meaning of an essential truth can be verified only to the extent language, sounds, or images can represent it, we may not know the actual degree of similarity between sets of experiences of multiple individuals. Nevertheless, as with most other types of information, we could assume a reasonable similarity level based on a subjective analysis of the individuals' utterances or different ways of communicating the essential truths of their respective experiences.

Valence as a momentary response to an experience naturally varies over time as the experience changes. Similar to the adage that one cannot step into the same river twice, the exact same experience cannot repeat itself. The universe, including humans, constantly changes. Besides, any experience will likely affect one's nervous system pathways and future perception [10].

Nevertheless, experiences over time can be the same at some level of universality if they share some pure essences. For example, one and the same ice hockey game constitutes one long experience at some level, albeit the details of that experience vary wildly throughout the game. At a more detailed level, we typically experience different things when the home team scores compared to when the visiting team scores. Consequently, we can refer to types of experiences to which we can attach additional pure essences that would alter the valence of that experience.

We can not fully perceive any situation as we cannot be aware of all its parts simultaneously. Hence, we cannot obtain the complete essence of a situation, and therefore valence can be attached only to a partial essence. For any type of experience, we can therefore expect the corresponding valence to fluctuate, resulting in a distribution of valence signals. The shape of that distribution would be contingent upon the set of possible instances with the partial essence of that type of experience in common.

Assuming valence to be a representation of value, we could, in theory, plot a value distribution over any type of experience. Each of the imaginable instances of a particular experience type could be assigned a more or less precise value corresponding to the valence of that imagined instance.

The older a valuation, the less impact it has on the valence of a current experience—eventually, it may have no impact at all. Value distributions, in this sense, are, therefore, dynamic. As a result, a valuation is the most precise only in the present moment where the pure essences of the current experience are as clear as possible. As soon as we generalize and thus move from the current to the imagined, we immediately have a set of possible value assignments that we, based on historical experience, could attach to possible instances of an imagined experience.

Suppose values are known or discovered through valence, and the nervous system functions similarly between individuals. In that case, humans would generally assign similar values to the same types of experiences. For example, we typically value the experience of eating when hungry based on the sense of taste and satiety, which creates a positive valence toward the experience of eating. On the other hand, the value of possessing a particular type of non-essential object could, for example, depend on culture and norms and possibly vary significantly between individuals of different backgrounds.

Assuming that all types of experiences result in a particular valence, we should be able to compare the value of experiences between individuals at some level of universality and under normal circumstances (i.e., in general or with reference to 'the typical imaginable instance' of a type of experience), not in the absolute sense but at least with regard to order and difference. An individual could, for example, value bicycling more than running. Such an utterance would not mean that all instances of bicycling are more valuable than running, but typically that is anticipated to be the case. Another individual could say the opposite, that running is more valuable than bicycling. Hence, the two individuals order bicycling and running differently based on value. In addition, the first individual could state that bicycling carries much more value than running, which is almost as valuable as walking, thereby expressing differences in value between the three activities.

Lastly, apart from the momentary experience, values can only be anticipated. Any value of an imagined future experience is, in a sense, expected rather than absolute. One of the reasons is the experience-induced plasticity of our nervous system [5]—similar situations could result in very different valence signals. Another reason is the impossibility of foreseeing the future precisely. Chances are that no actual experience will turn out exactly as imagined beforehand.

3 Anticipated and Actual Value Drivers

An anticipated value driver is a subset of the essential truths of the essence of an experience that has, as far as it is possible to determine, a noticeable impact on the valence of that experience. Not only can such essential truths determine the essence of the things present in a concrete instance of that type of experience. They can also decide the concrete historical essences [7] upon which any concrete instance of that type of experience is assumed to be contingent.

Value drivers are observable through valence. Since valence is a result of the whole of an experience, only experiences based on factual events can determine actual value drivers, for example, by removing them from or adding them to an actual experience. Furthermore, as the name implies, actual value drivers are based on real-world facts. Nevertheless, over time and with repeated occurrences of experiences that share substantial parts of their respective essences, the discrepancy between the essences of anticipated and actual value drivers should become less pronounced. Value drivers can be of various types, e.g., acts, objects, and conditions. Still, a value driver, anticipated or actual, is always a value driver *for* someone (or for a group as long as the group members have that value driver in common).

Value drivers can be causally related, and as such, a value driver can be more or less general. Money is an example of a rather general value-driver—as long as its essence consists of the possibility of it being exchanged for other goods—since it can be traded for more specific value drivers. Since, until the time of actual trade, money can only possibly be traded for something, and thus it would be an anticipated value-driver unless the experience of having the money has a positive effect on the valence of the current experience. Money could be used to buy a ticket, which itself would be an anticipated value-driver, albeit more specific, by possibly offering access to a particular ride at a theme park where the ride is part of a valuable experience. The more rides the ticket would be good for, the more general it would be as an anticipated value-driver.

While an anticipated value driver's existence is dependent on some person's essential insight, some actual value drivers might never be known. Certain situations are highly dependent on factors that we are quite unaware of, actual as well as historical. Sometimes the absence of something can make an experience great, even as we are rather unaware of that absence. Valence, as a property of experiences, is ever-present, regardless of our knowledge about what causes a particular level of it. Value drivers that, in hindsight, turn out to have been necessary for a valuable experience to occur may not be known at the time of their actual occurrence.

Some value drivers may, upon their realization, provide a rather unpleasant experience themselves. For instance, studying for and taking an exam may prove a grueling experience for some. Still, performing the acts of studying and exam writing are value drivers in an experience whose essence includes having passed an exam with flying colors. Similarly, some societies require prospective members to go through long periods of performing challenging tasks involving experiences to which a rather negative valence can be attached. Yet, as long as the experience of being an actual member is anticipated to come with a lasting positive valence, some will find taking on the challenging tasks worthwhile anticipated value drivers. Only when the positive valence of the experience of being a member of that society do the aforementioned value drivers become realized in the form of historical facts. It could, however, in retrospect, be looked upon and talked about as a valuable experience. In the latter case, however, it is probably the memory and the experience of having had the experience that are valuable. That memory could act as a constantly accessible value driver in the type of situations where it affects the value of the experience. Such an occasion could include community meetings, which would never have occurred should the admission process never have taken place. A bad experience can be anticipated to be, or prove to be, in hindsight, a value driver. The fact that you did it is always going to be present, and so a fact can be a value driver. As such, value drivers do not have to drive the value of an experience instantly. A value driver can possibly affect the value of future experiences.

Just as we cannot have the exact same experience more than once, and we, therefore, speak about the essence of experiences rather than actual ones, any given value driver cannot be realized multiple times. However, multiple instances of value drivers can, at some level of universality, have the same or similar essence. Therefore, in particular, when dealing with input from multiple stakeholders, we refer to the essences of value drivers, which at some level of universality, will suggest certain commonalities based on which conclusions about the values of the stakeholders can be drawn. Note that a value driver is not necessarily universally a value driver. Thus, an anticipated value-driver ought to be specified along with the essential experience of which it is a part.

Lastly, a mere change in time horizon may change the prospect of something being an anticipated value driver. For example, an individual at a bar may consider having another drink to constitute an anticipated value driver of some experience in the very near future. By extending the time horizon, however, the drink may no longer appear as much of a value driver as the value it is anticipated to contribute in the near future is contingent upon its anticipated contribution of a very negative value to some experience the following morning.

Attempts to find reasonable compromise solutions in participatory decision-making sometimes require the ability to generalize the value drivers collected from a group of stakeholders. While it may be notoriously difficult to satisfy a relatively detailed set of anticipated value drivers at once, the types of experiences they represent may be satisfiable at a more general level. Assume a number of stakeholders where each proposes his/her favorite sport as an anticipated value driver, and the set of favorite sports is quite large. Any direct agreement on a subset of the favorite sports would likely be difficult to reach. However, suppose the set of anticipated value drivers were generalized to sports without further specification. In that case, that generalization could then be used as a point of departure for further deliberation without disregarding the stakeholders' values.

Generalizing value drivers can be done based on the notion of types. Two value drivers are of the same type if and only if they both share the essence specifying that type. Types of value drivers can be pictured as a tree, with the root representing the most general type and the leaves the most specific. The structure of the tree will depend on the typology adopted.

A value-driver type is defined by a set of essential truths that the elements of that type satisfy. Let $\mathcal{V} = \{P_j\}$ be a value-driver type, where each P_j is an essential truth, then $\mathcal{V}' = \mathcal{V} \cup P'$, where P' is an essential truth distinct from all P_j, is a sub-type of $\mathcal{V}$, and $\mathcal{V}$ is a super-type of $\mathcal{V}'$—the addition of another essential truth increases the specificity and lowers the degree of universality. For example, a super-type could cover value drivers that provide warmth, and a sub-type of that super-type could specify warm jackets. Creating value-driver typologies delineates ways of going from the general to the more specific and vice versa.

4 Aggregating Value Drivers

A meaningful interpretation of collected value drivers depends on a mutual appreciation and shared understanding of the context between the respondent and the analyst. In general, however, a more general interpretation has a greater chance of capturing the meaning of the value driver than a very detailed interpretation. In the latter case, there are more possibilities for error. An aggregation based on value-driver types is advantageous for the analyst from the perspective of adequately representing meaning, in addition to what has been written above about the improved outlook of finding compromise solutions that seem reasonable to most stakeholders.

When collecting value drivers from stakeholders, it is imperative to provide the respondents with a context that is sufficiently clear and delineated, given the purpose of the inquiry. The analyst must not have to wonder what type of experience to which a value driver indeed was attached. Granted, the range of experiences a large number of respondents could imagine is vast. Nevertheless, a fruitful analysis and aggregation depend on a reasonable basis of commonalities between the essences of those imagined experiences.

Even though the essential being of one and the same thing may be different for different persons, it is realistic to assume that at least the majority and the most fundamental essential truths are true for almost everybody due to the common stock of knowledge at hand [12]. Some essential truths may not be part of the eidos of an object available to the consciousness of a blind person. To that individual, roundness may be an essential truth of the eidos of a soccer ball, but the black-and-white color pattern may not be. In any case, the reasonably common sets of essential truths of which valence is an inherent quality make it reasonable to speak about value-driver aggregation in the way we do here.

When aggregating value drivers, we need to consider reasonable levels of value-driver types as we seldom would be able to cater to each individual's value drivers specifically—we have to generalize. For example, suppose each individual wants a specific but warm jacket. In that case, we could generalize that into *a warm jacket* and then leave it to the decision maker to select the most suitable jacket. An even greater degree of generalization may reveal options that would keep the stakeholders warm in other ways that exclude warm jackets—if *staying warm*, for example, seems to be the most universal yet informative essence of the collected value drivers.

If there are conflicting value drivers, we need to consider the reason for obtaining them. Suppose the reason is to make as many people as possible happy. In that case, the solution might be to go with whatever makes the majority happy—albeit we realize that working out what makes the majority happy itself can be far from trivial. In any case, we should resolve conflicts based on some overarching principles, which need to account for the anticipated effect of catering to one set of value drivers over another.

Having suggested a particular kind of experience to a group of stakeholders, we assume that each stakeholder will have grasped the essence of that experience at some level. If one were to say, "Imagine yourself in a crowded stadium," one can assume that the receivers of that statement would, through ideation, obtain sets of essences that have certain essential truths in common. The stakeholders will not have the same intuition, but the intuitions will share enough essence (as long as the suggested experience is reasonably specified) for the stakeholders to be able to reason about that type of experience, even if none, in fact, experienced precisely the situation that was suggested—assuming that works, we can readily create a basis for collecting anticipated value-drivers of such a type of experience.

Partition the value drivers into sets $\Delta_1, \ldots$, such that the intersection of the value drivers in each set is nonempty—they must have some essential truths in common—and each set constitutes a meaningful value-driver type. Repeat the process for each set based on commonalities distinct from previous ones, i.e., if the intersection of the value drivers in Δ_i is $\mathcal{D}'$, then a partitioning of Δ_i must be done without regard to the elements in $\mathcal{D}'$. If further partitioning results in no relevant increase in specificity, the set constitute a *feasible* value-driver type. Partitioning Δ into Δ_1 and Δ_2 leads to a relevant increase in specificity if allocating resources based on Δ_1 and Δ_2 rather than on Δ improves the decision basis. A *strictly feasible* value-driver type contains at most one value driver per respondent. The partitions and sub-partitions are represented naturally as a tree. In general, one could generate multiple such trees for every set of value drivers. Furthermore, each tree would yield a different set of feasible or strictly feasible value-driver types—the decision maker will eventually have to select which should inform the final decision.

5 The Case of Kramfors' Growth Strategy

To better understand the values of the citizens and businesses of Kramfors and be able to incorporate those in the design of a municipal growth strategy, an online survey was distributed to 1080 citizens (including some municipal ambassadors who live elsewhere), and another similar survey to 750 businesses. The respondents were asked to list, at most, seven important factors associated with the municipality's development. For the citizens, the factors should be important for living and working in the municipality, while for the businesses, the factors should be important for a viable local commercial and industrial life. These important factors were expected to capture the seven most prominent anticipated value drivers for each respondent of the broad types of experiences: "You live and work in Kramfors," or "The business is part of a viable commercial and industrial life at the local level." The phrasings of the main questions were meant to set the corresponding contexts, i.e., the fundamental essences of the experiences suggested to the respondents. The citizens were asked "What do you find to be the most important factors for living and working in Kramfors?" and the business representatives were asked "What do you find to be the most important factors for a viable commercial and industrial life in Kramfors?" A follow-up question in the same survey asked each respondent to rate the factors he/she had listed in terms of how well they were fulfilled at the time of the survey on a five-grade scale from "not at all" to "very much." The samples are unlikely representative but should suffice for demonstrating the technique of aggregating value drivers.

The reason for collecting value drivers from citizens and businesses and creating corresponding typologies were twofold. Firstly, to get a better understanding of what type of strategy would make the municipality more palpable, thereby increasing the chance of current citizens staying in the municipality, as well as getting more individuals and families to settle there. Secondly, to see how the municipality could be made more attractive to various businesses, not least to increase the opportunities for citizens to get a job.

To create a value-driver typology, we proceeded as follows: (1) Partition the set of value drivers into subsets $\Delta_1, \Delta_2, \ldots$ such that for all $\mathcal{D}_k \in \Delta_i$, the intersection $\cup \mathcal{D}_k$ is non-empty, and at the current level of universality, all $\mathcal{D}_k$ are considered to be of the same value-driver type—what sets one value-driver type apart from another depends on the context as well as the person or group performing the analysis. (2) Continue partitioning $\Delta_1, \Delta_2, \ldots$ and their respective sub-partitions in the same manner until the following holds: (a) no value drivers in the current type have enough distinguishing features given the context such that another partitioning into more specific value-driver types is warranted, and (b) each value-driver type contains at most one value driver per respondent. The requirement (b) guarantees value-driver types that are specific enough to allow for interpretations such as "respondent a finds value-driver type Δ_i to be more important than Δ_j if and only if a stated the important factor (i.e., value driver) $\mathcal{D}_k \in \Delta_i$ to be more important than important factor $\mathcal{D}_l \in \Delta_j$."

The municipality selected a subset of the strictly feasible value-driver types based on their cardinality and the distribution of responses to the five-grade fulfillment rating to inform the strategy design process. The choice was qualitative and not based on rules or principles stipulated in advance.

The total number of collected value drivers for the citizen survey was 751. Those were categorized according to a typology with three levels. At the first level, there were 25 types. At the second level, 76, and at the third level, 94 types. Of these, 13 types at the third level were selected for further analysis—that corresponds to 7 of the 25 types at the first level. They were: job opportunities, housing, services, meaningful spare time, adequate infrastructure, schools, and welfare. The business survey resulted in 349 value drivers, with 20, 64, 79, and 81 value-driver types at levels 1, 2, 3, and 4, respectively. Eleven (11) of the types at level 1 were selected for further analysis: a responsive municipality, business development, adequate infrastructure, schools, supply of competence, networking and dialog, housing, services, entrepreneurial spirit, job opportunities, and education.

The level of specificity required to obtain strictly feasible value-driver types resulted in a poor representation of the commonalities between responses. Therefore, the municipality opted for types of a greater level of universality that would inform the remainder of the strategy process. Each of these types would subsequently be interpreted by considering their respective elements, i.e., the actual responses underlying each type. Taking the selected value-driver types into account supported the municipality in determining which challenges should be the strategy's focus. In addition, they are expected to be used in indicating a strategy's alignment with the values of the public.[1]

6 Discussion

The survey result provided the municipality with a better understanding of the anticipated value drivers of the citizens as well as the extent to which they are currently in place or otherwise observable. While some anticipated value drivers seem necessary to fulfill basic human needs, some are indeed dependent on the local environment. For example, the value drivers that make the living environment attractive can, in this case, be summarized as a house with a garden, close to nature in a beautiful setting, along with friendly people. We can easily picture another set of value drivers that would signify a rather different living environment in a major city, namely closeness to social activities, restaurants, and various cultural offerings. A survey like this can give a decision-maker an understanding of the value drivers of the local population. The survey results

[1] At the time of writing, this is a work in progress; the survey was carried out during the spring of 2021.

inform the design of the municipal growth strategy such that resources are allocated to effectively increase the value of living and working in the municipality in alignment with the citizens' responses.

As for the businesses, a suitable infrastructure was one of the more prominent value drivers. Of the 83 respondents, 34 found infrastructure to be important. While that observation alone does not tell the municipality much about how to allocate its resources effectively, such a finding provides the municipality with a reason to investigate the local businesses' infrastructure needs further. Even relatively high-level value drivers can point out critical areas for further inquiry.

The concepts and method presented herein can support group decisions based on value-focused thinking by initiating a participatory decision process with a focus on values rather than on attributes and weights—the latter comes later in the process and is out of this paper's scope. When trying to find feasible compromise solutions in participatory decision processes, it is essential to understand the stakeholders' values and use them as a basis for the fundamental objectives. Similarly, suppose a strategy is expected to allocate resources in accordance with the values of the public. In that case, those need to be collected and aggregated in a meaningful way to inform the strategy design process before generating particular strategy alternatives.

7 Conclusion

We have proposed and discussed the concept of value drivers as fundamental in participatory group decision analysis. In particular, for situations where a decision-making body is interested in gaining a better understanding of the *values of the public*. It is typical for decision analyses within the public domain to use criteria to represent the administration's idea of public values—these are likely to be less connected to the values of the public within the decision context. However, criteria impose certain requirements, such as how to express preferences over different outcomes and, in some cases, indicators expressing performance levels. With conflicting objectives, means for weighting criteria against each other are needed. Eliciting attributes and weights directly from a large group of disparate stakeholders is bound to be complicated. Although value drivers cannot substitute criteria, they can motivate the development of criteria as representations of the values of the public in a given context. The qualitative nature of the method, however, limits the number of respondents; we will attempt to address that in future studies. In conclusion, for conventional participatory decision analysis with multiple objectives, value drivers can form a basis for decision criteria, thereby acknowledging the values of the public and facilitating the incorporation of the citizens' worldview and concerns in a public and participatory decision process.

Acknowledgements. The authors would like to thank Andreas Gylling at the Municipality of Kramfors, IIASA, and the anonymous reviewers.

References

1. Bayley, C., French, S.: Designing a participatory process for stakeholder involvement in a societal decision. Group Decis. Negot. **17**, 195–210 (2008)
2. Carruthers, P.: Valence and value. Philos. Phenomenological Res. **97**(3), 658–680 (2018)
3. Choo, E.U., Schoner, B., Wedley, W.C.: Interpretation of criteria weights in multicriteria decision making. Comput. Ind. Eng. **37**(3), 527–541 (1999)
4. Earl, G., Clift, R.: Stakeholder value analysis: a methodology for intergrating stakeholder values into corporate enviromental investment decisions. Bus. Strategy Environ. **8**(3), 149–162 (1999)
5. Galván, A.: Neural plasticity of development and learning. Hum. Brain Mapp. **31**(6), 879–890 (2010)
6. Husserl, E.: Ideas: General Introduction to Pure Phenomenology. Routledge, Milton Park (2012)
7. Kaufmann, F.: The phenomenological approach to history. Philos. Phenomenological Res. **2**(2), 159–172 (1941)
8. Keeney, R.L.: Value-Focused Thinking: A Path to Creative Decisionmaking. Harvard University Press, Cambridge (1992)
9. Kong, M.S., Zweifel, L.S.: Central amygdala circuits in valence and salience processing. Behav. Brain Res. **410**, 113355 (2021)
10. Morrison, S.E., Salzman, C.D.: Re-valuing the amygdala. Curr. Opin. Neurobiol. **20**(2), 221–230 (2010)
11. Ortiz, G., Domínguez-Gómez, J.A., Aledo, A., Urgeghe, A.M.: Participatory multicriteria decision analysis for prioritizing impacts in environmental and social impact assessments. Sustain.: Sci. Pract. Policy **14**(1), 6–21 (2018)
12. Schutz, A.: Husserl's importance for the social sciences. In: Collected Papers I, pp. 140–149. Springer, Cham (1962)

Conflict Modeling and Distributive Mechanisms

Composite Consensus-Building Process: Permissible Meeting Analysis and Compromise Choice Exploration

Yasuhiro Asa[1(✉)], Takeshi Kato[2], and Ryuji Mine[1]

[1] Hitachi Kyoto University Laboratory, Hitachi Ltd., Kyoto 606-8501, Japan
yasuhiro.asa.mk@hitachi.com
[2] Hitachi Kyoto University Laboratory, Kyoto University, Kyoto 606-8501, Japan

Abstract. To solve today's social issues, it is necessary to determine solutions that are acceptable to all stakeholders and collaborate to apply them. The conventional technology of "permissive meeting analysis" derives a consensusable choice that falls within everyone's permissible range through mathematical analyses; however, it tends to be biased toward the majority in a group, making it difficult to reach a consensus when a conflict arises. To support consensus building (defined here as an acceptable compromise that not everyone rejects), we developed a composite consensus-building process. The developed process addresses this issue by combining permissible meeting analysis with a new "compromise choice-exploration" technology, which presents a consensusable choice that emphasizes fairness and equality among everyone when permissible meeting analysis fails to do so. When both permissible meeting analysis and compromise choice exploration do not arrive at a consensus, a facility is provided to create a sublated choice among those provided by them. The trial experimental results confirmed that permissive meeting analysis and compromise choice exploration are sufficiently useful for deriving consensusable choices. Furthermore, we found that compromise choice exploration is characterized by its ability to derive choices that control the balance between compromise and fairness. Our proposed composite consensus-building approach could be applied in a wide range of situations, from local issues in municipalities and communities to international issues such as environmental protection and human-rights issues. It could also aid in developing digital democracy and platform cooperativism.

Keywords: Consensus Building · Discussion Support · Conflict Resolution

1 Introduction

Today's social issues are exacerbated because of social conflicts caused by diverse values, such as order formation in international conflicts, environmental conservation such as global-warming countermeasures, and human-rights issues such as well-being and gender equality. Otherwise expressed, social issues are deeply related to problems arising from conflicting opinions in groups although there may be many opinions in such groups, and group decision-making methods are needed to solve these problems.

Y. Maemura et al. (Eds.): GDN 2023, LNBIP 478, pp. 97–112, 2023.
https://doi.org/10.1007/978-3-031-33780-2_7

Typically, in the real world, voting via preference aggregation rules is widely used for group decision-making. Various types of preference aggregation rules exist, including simple majority voting, scoring rules such as the Borda count, and cumulative voting [1, 2]. However, these preference aggregation rules can cause cyclical preferences and can make consensus impossible (Condorcet's paradox) [3], and it has been proven that the conditions of fairness, Pareto efficiency (unanimity), completeness and transitivity of preference relations, independence of choices, and non-dictatorship, cannot be satisfied simultaneously (Arrow's impossibility theorem); therefore, it is not possible to make unique and fair decisions by voting [4].

Social-choice theory is an academic field for social group decision-making that is desirable for all society members. One of the philosophies in this theory is "the greatest happiness of the greatest number (utilitarian principle)" proposed by the philosopher Bentham, which considers the relief of many poor people in a society of inequality as justice [5]. By contrast, the philosopher Rawls proposed that it is just to make the disparate society itself, which Bentham assumed equal and fair, and advocated "maximization of the benefits of the most disadvantaged (the Difference Principle)" [6]. Group decision-making based on Bentham's principle, like the voting described above, tends to bias the result toward the majority. Conversely, although Rawls' principles value diversity and fairness and respect for disadvantaged minorities, they are difficult to realize through voting.

Such group decision-making methods that emphasize individual freedom, equality, and fairness include discussion and deliberation. The social philosopher Habermas advocated the importance of group decision-making that emphasizes intersubjectivity through communicative acts in discussion [7]. In addition, the political scientist Fishkin has described the importance of decision-making based on changes in participants' opinions before and after deliberation through deliberative polling [8]. The political scientist Gutmann proposed deliberative democracy in which people listen to the opinions of others as a form of democracy through discussion [9], and the necessity and practicality of deliberative democracy in the field of education and urban development have been examined in Japan as well [10, 11].

The process of group decision-making through discussion and deliberation is important. Social psychologists Thibaut and Walker have shown that free discussion among participants can lead to satisfaction with the outcome and a sense of fairness [12], and psychologist Leventhal has described the factors in the process that promote a sense of fairness (procedural fairness criteria) [13]. The anthropologist Graeber has stated that original democracy is "a process of compromise and synthesis in which no one goes so far as to refuse to agree" [14]. Following Graeber's argument, "consensus" in this study is defined as an acceptable compromise that not everyone rejects. Specific methods for such a process include the spokes council in which spokes (representatives) are determined in units of affinity groups and discussions are held [15], and the consensus-building method advocated by the urban planner Susskind [16].

To support the process of group decision-making through discussions, several technologies with online opinion aggregation and chat functions have been proposed. For example, Decidim [17] has a full range of auxiliary components, such as questionnaires and blogs, vTaiwan [18] automatically groups participants' opinions, Loomio [19] and

Liqlid [20] visualize opinions through pie charts and word clouds, and D-Agree [21] has functions for automatically structuring opinions and facilitation via artificial intelligence (AI). However, although these technologies are expected to be highly effective in activating and organizing opinions, they do not have a support function when opinions conflict, and it is difficult for participants to reach a consensus, as Graeber [14] advocates.

A mathematical framework for resolving conflicts of opinion is the graph model for conflict resolution (GMCR) [22–28]. The GMCR expresses the structure of conflicting opinions in a graphical model and performs mathematical analysis based on the preference order of decision makers for each opinion to derive a consensusable choice that takes rationality and efficiency into account. However, there is a problem that the number of states to be analyzed is as huge as A^N (A: number of choices, N: number of decision makers; computational complexity order $O(k^n)$, , exponential time). Therefore, permissible meeting analysis (PMA) has been proposed to avoid the problem, referring to the idea of GMCR [29, 30]. This method derives a consensusable choice that considers permissibility by performing a mathematical analysis based on ordering the decision makers' preferences for the choices and their permissibility. The number of states to be analyzed is practical as it is the same as the number of choices A (computational complexity order $O(n)$, , linear time). However, as PMA gives priority to the consensusable choice with the smallest total adjustment when adjusting everyone's permissible range, it is easy to derive a consensusable choice that is less burdensome for the majority, and it may be difficult to reach a consensus, including the minority group.

Based on the above discussion, our objective was to provide a new consensus-building support technology to reach a consensus and a new consensus-building process for using this technology. Specifically, in addition to the conventional technology PMA, we provide a new technology—compromise choice exploration (CCE). A composite consensus-building process that combines the two consensusable choices derived from both supports a consensus. As mentioned, the problem is that PMA alone tends to present a consensusable choice that is biased toward the majority. Therefore, we provide a technology for presenting a consensusable choice that emphasizes fairness, where compromise among all is equalized by CCE when PMA fails to promote consensus. The new composite consensus-building process combines PMA and CCE to enable a consensus that falls within everyone's permissibility and range of compromise while emphasizing fairness. These technologies and processes expand the research streams of Rawls' principle of difference, the deliberative democracy of Gutmann et al., Leventhal's procedural justice, and Graeber's compromise and synthesis, and they put forth new research directions in terms of developing conventional opinion exchange tools and mathematical analysis models.

2 Methods

2.1 Permissible Meeting Analysis

First, we explain the conventional conflict-resolution technology, PMA, as our composite consensus-building process employs PMA as one of its functions.

In a discussion, let $M = \{1, 2, \cdots, m\}$ be the set of participants, $X = \{x_1, x_2, \cdots, x_n\}$ be the set of choices, and $\succsim_i = x_{i1} \succ x_{i2} \succ \cdots \succ x_{in}$ be the preference order for the choices of participant $i \in M$. Let $x_{ij} \in X, j$ be the preference order, $maxP_i = \{x_{i1}, x_{i2}, \cdots, x_{ik_i} | k_i \leq n\}$ be the set of acceptable choices for participant i, and $maxP_i^l = \{x_{i1}, x_{i2}, \cdots, x_{ik_i}, \cdots, x_{i(k_i+l)} | (k_i + l) \leq n\}$ be the set of acceptable choices for $maxP_i$ extended by l.

The algorithm for PMA is shown in Fig. 1. In PMA, consensusable choices are those that exist within the permissible range of all participants. Therefore, in Step 1, the product set U_o of acceptable choices for participant $i(= 1, 2, \cdots, m)$ is derived using Eq. 1, and whether $U_o = \varnothing$ is determined. If $U_o = \varnothing$ does not hold, then the process moves to Step 2, outputs the set of alternatives U_o derived in Eq. 1 as the consensusable choice X_o, and terminates.

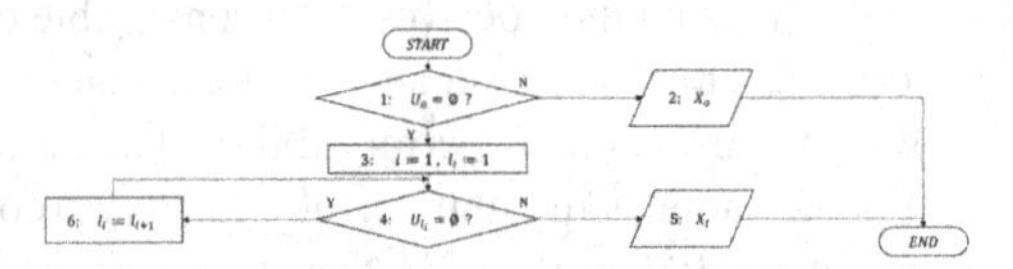

Fig. 1. Flowchart of permissible meeting analysis

If $U_o = \varnothing$ in Step 1, the permissible range l_i of each participant i is expanded individually in the loop of Steps 3, 4, and 6 to derive the union set U_{l_i} of the permissible ranges of participant $i(= 1, 2, \cdots, m)$, respectively, as shown in Eq. 2, and to determine whether $U_{l_i} = \varnothing$. If $U_{l_i} = \varnothing$ is no longer the case in Step 4, then in Step 5, the consensusable choice is the set of choices in the product set U_{l_i} that has the smallest permissible range of l_i, i.e., the set X_l of choices that is the smallest in Eq. 3.

$$U_o = \bigcap\nolimits_{i=1}^{m} maxP_i \tag{1}$$

$$U_{l_i} = \bigcap\nolimits_{i=1}^{m} maxP_i^{l_i} (l_i = 0, \cdots, n - k_i) \tag{2}$$

$$X_l = \underset{U_{l_i}}{\operatorname{argmin}} \sum\nolimits_{i=1}^{m} l_i. \tag{3}$$

If the set X_o in Step 2 or the set X_l in Step 5 contains more than one consensusable choice, all of them shall be considered as the result of the derivation from PMA. Note that the loop process in Steps 3, 4, and 6 does not fall into an infinite loop because there is always a permissible range l_i such that $U_{l_i} \neq \varnothing$. For reference, see Sect. 3.1 of [33] for an example of PMA.

2.2 Compromise Choice Exploration

Based on the PMA described in the previous section, we propose our new consensusable choice derivation technology, CCE. In PMA, the participants' act of compromise is to expand the permissibility range for the preferential order of choices, and as a consensusable choice is derived that minimizes the overall act of compromise, the compromise is likely to be biased toward the minority rather than the majority, resulting in a lack of fairness. Therefore, in CCE, the participants' act of replacement ordering for their preference order is regarded as a compromise act by the participants, and a common preference order that makes the number of replacement operations by each participant as equal as possible is regarded as a consensusable choice. This will lead to a highly fair consensusable choice.

The algorithm for CCE is shown in Fig. 2. CCE searches for a preference order among the list $(\succsim)_{all}$ of all preference orders for choice X, such that the number of order replacements from the initial preference order $\succsim_i$ for each participant i is equal. First, Step 1 generates $n!$ lists $(\succsim)_{all}$. Next, to calculate the number of replacements between the preference order $(\succsim_i)_{i\in M}$ of each participant i and each preference order of $(\succsim)_{all}$ for this list, in Step 3, as in Eq. 4, each element of the preference order of participant i is replaced by a number in ascending order according to the replacement rule $Rule_i$ in ascending numerical order, as shown in Eq. 4. Where $Rule_i$ means the replacement of x_{i1} with 1, x_{i2} with 2, x_{i3} with 3, and x_{in} with n.

$$Rule_i = \begin{pmatrix} x_{i1} \; x_{i2} \cdots x_{in} \\ 1 \;\; 2 \cdots n \end{pmatrix}. \tag{4}$$

In Step 5, we replace each preference order in the preference-order list $(\succsim)_{all}$ with the same rule $Rule_i$ as in Eq. 4, as in Eq. 5. Where $j = 1, \cdots, n!$.

$$\succsim_j' = \begin{pmatrix} x_{j1} \; x_{j2} \cdots x_{jn} \\ x_{j1}' \; x_{j2}' \cdots x_{jn}' \end{pmatrix}. \tag{5}$$

Let $Sort(x_{i1}, x_{i2}, \cdots, x_{in})$ denote the process of sorting any preference-order $(\succsim_i)_{i\in M}$ in ascending order by bubble sort, and $SortCount(x_{i1}, x_{i2}, \cdots, x_{in})$ denote the process of finding the number of sorting at that time [34]. In Step 6, each preference order in the preference-order list $(\succsim)_{all}'$ replaced by Eq. 5 is sorted in ascending order, as shown in Eq. 6, and the number of sortings r_{js} at that time is calculated according to Eq. 7.

$$\succsim_{js}' = Sort\left(x_{j1}', x_{j2}', \cdots, x_{jn}'\right) \tag{6}$$

$$r_{js} = SortCount\left(x_{j1}', x_{j2}', \cdots, x_{jn}'\right) \tag{7}$$

In the loops of Steps 7 and 8, and Steps 9 and 10, the process in Steps 3, 5, and 6 is performed for all participants i and all choices j.

In Step 11, based on the number of replacements for each list in the preference-order list $(\succsim)_{all}'$ obtained from each participant's preference order $\succsim_i$, we search for the one among $(\succsim)_{all}'$ for which the number of replacements for all participants is equal. Specifically, based on Eqs. 8 and 9, the *Average* μ and *Standard deviation* σ of the number of replacements $r_i (i = 1, \cdots, m)$ for each participant i for each list of $(\succsim)'_{all}$ are derived and *Score* is calculated.

$$\sigma = \sqrt{\frac{1}{m} \sum_{i=1}^{m} (r_i - \mu)^2} \tag{8}$$

$$Score = \mu + \sigma. \tag{9}$$

In Step 12, the list of preference order that minimizes *Score* in Eq. 10 is selected. The *Average* μ is smaller the less the degree of compromise from the initial preference order. *Standard deviation* σ is smaller for lists where all participants have the same number of replacements (when the value is 0, all participants have the same number of replacements). Therefore, Eq. 9 has smaller *Score* when the degree of compromise by each participant is as small as possible and as fair as possible, and Eq. 10 derives the option with the smallest *Score* as the most consensusable choice.

$$X_s = \underset{(\succsim)'_{all}}{\operatorname{argmin}}\, Score. \tag{10}$$

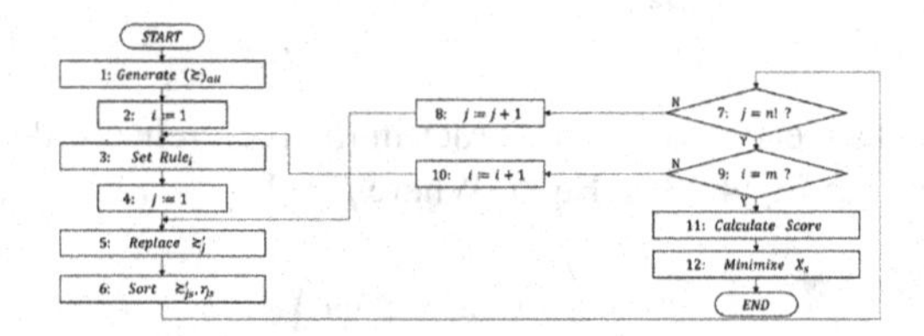

Fig. 2. Flowchart of compromise choice exploration

Note that, because the number of $(\succsim)_{all}$ to be fully searched in CCE is $n!$ (the computational complexity order is $O(n!)$, factorial time), the computational load increases as the number of choices increases. However, the number of choices managed in actual consensus building is approximately 10 at most, and a full search is not considered problematic for practical use. Although we used a full search here, a more efficient algorithm that gradually expands the replacement network from the participant's initial preference order and searches for the option with the smallest *Score* could be considered instead. For reference, see Sect. 3.2 of [33] for an example of CCE. There is also the Kemeny-Young method, a conventional method that uses each participant's preference order, but it is a method that derives a preference order that is close to everyone's preference by pairwise comparison counting, and is not a method that makes the number of preference order replacements as equal as possible, as in CCE [35].

2.3 Composite Consensus-Building Process

In addition to the PMA and CCE described so far, we propose a new composite consensus-building process that combines the PMA and CCE with sublated choice creation (SCC), which in turn creates a new sublated choice from multiple choices derived from the former two. Figure 3 depicts the flow of the consensus-building process for the problem of "deciding on the option that all participants agree from multiple choices."

First, in Step 1, PMA is performed based on the algorithm described in Sect. 3.1, and, based on the results, in Step 2, the participants discuss whether consensus is possible. In this discussion, if the same choices exist within everyone's permissible range from the beginning and a consensus is reached, then the discussion may be terminated. If the same choice does not exist, the consensusable choice derived in the Step 1 PMA and the permissible-range conditions at that time are presented to the participants, and if consensus is reached among all participants, the process is terminated.

If no consensus is reached among all participants in Step 2, they proceed to Step 3 and perform CCE based on the algorithm described in Sect. 3.2. Because the nature of the consensusable choices derived by PMA tends to be biased in favor of the majority, if consensus is not reached in Step 2, attempted fairness to the participants is considered to be the cause. The consensusable choice derived in CCE is the fairer choice and is more likely to achieve consensus. Therefore, in Step 4, the consensusable choice derived in CCE is presented to the participants to discuss the possibility of consensus. If consensus can be reached in this discussion, the process will be terminated.

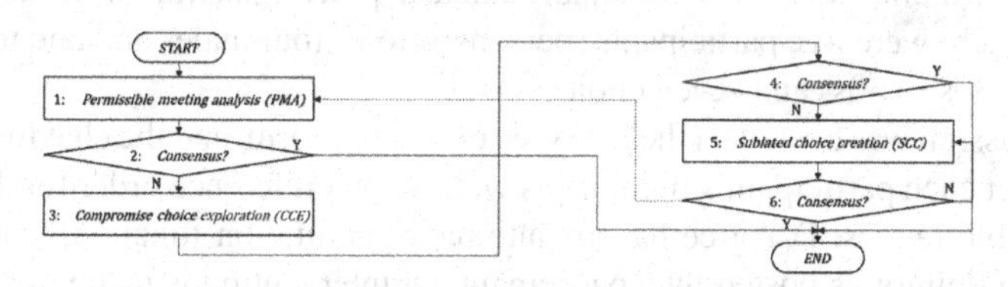

Fig. 3. Proposed composite consensus-building process

If no consensus is reached in Step 4, proceed to Step 5 to perform a new SCC based on the results of Step 1 (PMA) and Step 3 (CCE). Specifically, the contents of the choices derived by PMA that are likely to fall within everyone's permissibility range and the contents of the choices derived by CCE with high-rank preference order based on fairness are presented to the participants, and a new choice (sublated choice) is created by combining these contents. Then, in Step 6, consensusability will be discussed based on this sublated choice. If consensus is reached in this discussion and a consensus choice is determined, the process is terminated. The specific algorithm for SCC (PMA and CCE synthesis method) needs to be studied in the future. Tips for the SCC hints are discussed below in the last paragraph of Sect. 3.2.

If Step 6 still does not produce a consensus, return to Step 5, increase the number of consensusable choices derived from PMA and CCE, and recreate a new sublated choice. Specifically, PMA uses the choice with the smallest Eq. 3 and the second-smallest choice, and CCE uses the choice with the smallest Eq. 9 and the second-smallest choice, thereby creating a new sublated choice that combines them. Then, based on the sublated choice, the discussion is held again in Step 6.

Here, if multiple recreated sublated choices are created, it is possible to return to the Step 1 process as PMA and CCE are possible depending on the preference order and permissible range of the participants for these multiple sublated choices.

Thus, in our newly proposed composite consensus-building process, aiming for consensus, we first use PMA for a discussion based on a consensusable choice that falls within everyone's permissible range, and if consensus cannot be reached, we use CCE for a discussion based on a consensusable choice that considers fairness. Then, if consensus cannot be reached using either, a sublated choice is created by synthesizing the consensusable choices of PMA and CCE and is discussed. Conventional processes allow no effective means of supporting participants to reach a consensus when there are conflicts of opinion, but this process can support facilitation by presenting an effective consensusable choice.

3 Results

3.1 Trial Setup

To evaluate our proposed consensus-building support technology, we conducted trial experiments on PMA and CCE under the theme presented in Table 1. As there is no standard benchmark dataset on consensus building, we developed our own dataset in this study. The theme was "How to handle nuclear power generation in Japan for the future," and there were five participants (persons A to E: four males and one female; age ranging from 30s to 60s) and seven choices.

The discussion was hosted on the D-Agree online platform, and the chat function was used to collect each participant's opinion, as well as the preference order for the choices and permissible ranges. D-Agree has an automatic facilitation function, which means that, when an opinion is posted by a participant, an intervention is made in response to that opinion to encourage others to contribute, thereby stimulating discussion.

3.2 Trial Results

Table 2 shows the results of obtaining the preference order for the choices and permissible ranges for the seven choices of the five participants for the theme in Table 1.

PMA was conducted for the results presented in Table 2 according to the method described in Sect. 3.1. The resulting consensusable choice was (4) "no new nuclear power plants but restarting nuclear power plants is possible until alternative power generation methods are established," which was within the permissible range for all the participants. In addition, see Sect. 4.2 of [33] for the results of CCE performed on the results presented in Table 2 according to the method described in Sect. 2.2 of the present paper.

As mentioned earlier, within the participants' preference order and permissible range shown in Table 2, choice (4) was present in everyone's permissible range from the beginning, choice (4) chosen by PMA coincided with choice (4) chosen by CCE, and there was no conflict of opinion state. Therefore, to evaluate a case where there is a conflict of opinions, we assume, for example, as shown in Table 3, that there is no consensus choice within the permissible range of all the participants. This corresponds to the process from Step 1 to Steps 3 and 4 in Fig. 1.

Table 4 shows the results of PMA for the assumptions in Table 3. In this case, choice (1) "zero nuclear power plants by 2030," which has the smallest value of 2 for $\sum_{i=1}^{5} l_i$,

Table 1. Consensus building: Theme and choices

Theme	Discussion	Choices
Nuclear Power Generation	How should nuclear power be managed in Japan's future energy policy?	(1) Zero nuclear power plants by 2030 (2) Nationalize and decommission nuclear power plants (3) No new nuclear power plants; but restarting nuclear power plants is possible on the condition of safety and local consent (4) No new nuclear power plants; but restarting nuclear power plants is possible until alternative power generation methods are established (5) Restart nuclear power plants intending to decommission them and promote the development of next-generation nuclear power plants (6) Nuclear power plants can be operated with emphasis on safety (7) Proactively utilize nuclear power plants

Table 2. Preference order and permissible range (white: permitting, gray: not permitting)

	"A"	"B"	"C"	"D"	"E"
Rank 1	(5)	(4)	(4)	(5)	(4)
Rank 2	(4)	(3)	(3)	(4)	(2)
Rank 3	(3)	(2)	(2)	(3)	(1)
Rank 4	(2)	(6)	(1)	(2)	(3)
Rank 5	(1)	(1)	(5)	(1)	(5)
Rank 6	(6)	(7)	(6)	(6)	(6)
Rank 7	(7)	(5)	(7)	(7)	(7)

which represents the degree of widening of the overall permissible range, is derived as the consensusable choice, and the permissible range conditions at that time are to widen the permissible ranges of two persons, B and D, from 4 to 5. Although the examples in Table 3 are few, this suggests that persons A, C, and E do not need a wider permissible range and that PMA tends to be biased in favor of the majority.

Table 3. Preference order and permissible range (white: permitting, gray: not permitting)

	"A"	"B"	"C"	"D"	"E"
Rank 1	(5)	(4)	(7)	(5)	(6)
Rank 2	(4)	(3)	(6)	(4)	(7)
Rank 3	(3)	(2)	(2)	(3)	(1)
Rank 4	(2)	(6)	(1)	(2)	(5)
Rank 5	(1)	(1)	(4)	(1)	(3)
Rank 6	(6)	(7)	(3)	(6)	(4)
Rank 7	(7)	(5)	(5)	(7)	(2)

Note that Table 4 shows that there are three cases where the value of $\sum_{i=1}^{5} l_i$ is 3, and their consensusable choices are different from (2), (4), and (3). This suggests that although one consensusable choice could be selected in the hypothetical example in Table 3, in some situations, there are cases in which the value of $\sum_{i=1}^{5} l_i$ alone does not narrow down to a single consensusable choice. This is discussed below in the "Discussion" section.

Table 5 shows the results of CCE performed on the results presented in Table 3. The preference order {(4),(6),(7),(5),(3),(2),(1)} that minimizes *Score* and the first-ranked choice (4) "no new nuclear power plants, but nuclear power plants can be restarted until alternative power generation methods are established" are derived as consensusable choices. However, Table 5 shows several cases in which *Score* and *Standard deviation* σ have the same value. This suggests that, although one consensusable choice could be selected in the assumed example in Table 3, in some cases, *Score* or *Standard deviation* σ alone may be insufficient to narrow the consensus to one consensusable choice. This is discussed below in the "Discussion" section.

Unfortunately, no consensus was reached in the results of this study, but the following useful findings were obtained. In the assumed example in Table 3, choice (1) derived by PMA did not match choice (4) derived by CCE. This is because PMA is based on permissible-range expansion and CCE is based on preference-order replacement. In CCE, the first preference order choice is meaningful and the high-rank choices are meaningful as a compromise range.

Therefore, if we rearrange the first-place choice (4) in the CCE, the second-place choice (6), and the PMA choice (1), we get "(4) no new nuclear power plants, but nuclear power plants can be restarted until alternative power generation methods are established, (6) nuclear power plants can be operated with the emphasis on safety, and (1) zero nuclear power plants by 2030." By synthesizing choices (4), (6), and (1), the choice "no new nuclear power plants, but restarting nuclear power plants is possible with the emphasis on safety until alternative power generation methods are established, to attain zero nuclear power plants by 2030" emerges. Therefore, choices (4), (6), and (1) do not conflict but suggest a sublated choice that combines them. Otherwise expressed, a consensus process that combines PMA and CCE allows for an eclectic mix of both outcomes to reach a sublation.

Table 4. Results of the permissible meeting analysis for the assumptions listed in Table 3

Permissible Range						$maxP_A^{l_A}$	$maxP_B^{l_B}$	$maxP_C^{l_C}$	$maxP_D^{l_D}$	$maxP_E^{l_E}$	$\bigcap_{i=1}^{5} maxP_i^{l_i}$
k_A+l	k_B+l	k_C+l	k_D+l	k_E+l	$\sum_{i=1}^{5} l$						
5	5	4	5	4	2	(5)(4)(3)(2)(1)	(4)(3)(2)(6)(1)	(7)(6)(2)(1)	(5)(4)(3)(2)(1)	(6)(7)(1)(5)	(1)
5	4	4	4	7	3	(5)(4)(3)(2)(1)	(4)(3)(2)(6)	(7)(6)(2)(1)	(5)(4)(3)(2)	(6)(7)(1)(5)(3)(4)(2)	(2)
5	4	5	4	6	3	(5)(4)(3)(2)(1)	(4)(3)(2)(6)	(7)(6)(2)(1)(4)	(5)(4)(3)(2)	(6)(7)(1)(5)(3)(4)	(4)
5	4	6	4	5	3	(5)(4)(3)(2)(1)	(4)(3)(2)(6)	(7)(6)(2)(1)(4)(3)	(5)(4)(3)(2)	(6)(7)(1)(5)(3)	(3)
⋮	⋮	⋮	⋮	⋮	⋮	⋮	⋮	⋮	⋮	⋮	⋮

Table 5. Results of the compromise choice exploration for Table 3

$(\precsim)_{all}$	r_{js} for $\precsim'_A$	r_{js} for $\precsim'_B$	r_{js} for $\precsim'_C$	r_{js} for $\precsim'_D$	r_{js} for $\precsim'_E$	Average "μ"	Standard deviation "σ"	Score "$\mu + \sigma$"
{(4),(6),(7),(5),(3),(2),(1)}	9	8	10	9	8	8.8	0.748	9.548
{(4),(6),(7),(5),(2),(3),(1)}	10	9	9	10	9	9.4	0.49	9.89
{(4),(7),(6),(5),(3),(2),(1)}	10	9	9	10	9	9.4	0.49	9.89
{(4),(6),(5),(2),(7),(1),(3)}	9	10	10	9	10	9.6	0.49	10.09
{(6),(4),(5),(2),(7),(3),(1)}	9	10	10	9	10	9.6	0.49	10.09
{(6),(4),(5),(2),(1),(7),(3)}	9	10	10	9	10	9.6	0.49	10.09
{(4),(7),(5),(6),(3),(2),(1)}	9	10	10	9	10	9.6	0.49	10.09
⋮	⋮	⋮	⋮	⋮	⋮	⋮	⋮	⋮

4 Discussion

While conventional discussion-support technologies have difficulty in aggregating opinions, and GMCR, a conventional conflict resolution technology, is burdened with huge computational complexity (computational complexity: number of choices to the power of the number of participants), PMA (computational complexity: number of choices), a conflict-resolution technology, and CCE (computational complexity: factorial of the number of choices), a newly proposed technology, can support facilitation by presenting agreeable proposals toward consensus of all members while reducing computational complexity.

In PMA, as shown in Table 4, a consensus is reached by selecting a choice that falls within the permissible ranges of all participants. However, as this consensusable choice minimizes the extent of the permissible range of all participants, the consensus it produces tends to be biased toward the majority.

In CCE, as shown in Table 5, by selecting the choice with the smallest *Score*, a consensusable choice is derived with the fewest possible compromises for each participant and with a high degree of fairness. In Eq. 9, which was used to calculate *Score*, the *Average* μ and *Standard deviation* σ are added with the same weighting factor, but if it is desirable to reduce compromise and emphasize the majority, increase the factor of μ, and to respect the minority and emphasize fairness, increase the factor of σ. Thus, the weighting of compromise and fairness can be varied depending on the social issue.

If consensus is not reached as shown in Table 4 and Table 5 through steps 1 through 4 of the composite consensus-building process (Fig. 3), an auxiliary step between Steps 4 and 5 may be proposed. In this auxiliary step, the PMA and CCE consensus choices are compared, and if there is overlap between them, the choices are narrowed down and presented again to the participants to discuss the pros and cons of consensus. If there is no overlap between them, then it is necessary to proceed to Step 5 and derive a new sublated choice based on the content of both the PMA and CCE options, as described in the last paragraph of Sect. 3.2.

A composite consensus-building process that combines PMA and CCE can provide consensusable choices that fall within the permissibility range of all participants or that balance the number of replacements (compromise) and fairness. If consensus is not reached by PMA, it can proceed to CCE, and if consensus is still not reached, it can facilitate the creation of a new sublated choice from both PMA's and CCE's consensusable choices.

5 Conclusions

We developed a new compromise choice-exploration technology aimed at achieving consensus and provided a new composite consensus-building process that combines PMA and CCE. We conducted a trial experiment according to this process, and, based on the results, we obtained the following findings:

- We confirmed that PMA would derive a choice that was within the permissible range of all participants that would lead to consensus. However, it was found that consensusable choices that were biased toward the majority tended to be derived because the choice that minimized the permissible range of all the participants was prioritized.
- We confirmed that CCE derives a consensusable choice for participants with the fewest possible compromises and with high fairness by deriving a preference order that minimizes *Score*, which is the sum of the *Average* μ and *Standard deviation* σ of the number of replacements in the preference order. We also found that the weight coefficients of the *Average* μ and *Standard deviation* σ can change the balance between the overall number of replacements (compromise) and fairness.
- The composite consensus-building process can provide a consensusable choice that emphasizes fairness, where CCE provides an equal degree of compromise for all if PMA does not lead to consensus. Even if consensus cannot be reached using both, a sublated choice can be obtained via SCC, which synthesizes the consensusable choices of PMA and CCE.

Based on this trial experiment, it was found that the priority and overlap of choices should be considered when it is not possible to reduce to a single consensusable choice in PMA or CCE. It was also necessary for CCE to change the weighting of the *Average* number of replacements (compromise) and standard deviation (fairness) of the *Score* depending on the social issue. In addition, it would be desirable to work on technology to create a sublated choice in the future for cases where a consensusable choice cannot be found in either PMA or CCE.

The proposed consensus process consists of PMA and CCE; therefore, in the future, it will be combined with an online discussion platform that has a series of functions such as proposal, discussion, facilitation, and decision. Although this trial experiment was conducted on a small scale, which represents a study limitation, we intend to conduct controlled experiments and fieldwork on a statistically significant scale targeting municipalities and local communities to put this method to practical use as a group decision-making method for solving social problems.

In the future, our proposed approach can be applied to a wide range of practical situations, from local issues in municipalities and communities to international issues such as environmental protection and human rights issues. It could also aid in the development of digital democracy [31] and platform cooperativism [32]. In practical use, a large number of choices may be disadvantageous because of the time required to characterize respondents and search in the CCE, but the platform can avoid practical problems by narrowing down the choices through pre-discussion and multi-leveled processes. The former cites freedom and equality using digital technology, while the latter refers to joint ownership and the democratic governance of information platforms. We believe that the composite consensus-building process presented in this study will contribute to these movements.

Acknowledgments. The research for this study was conducted collaboratively between the Tokyo Institute of Technology (Tokyo Tech) and the Hitachi Kyoto University Laboratory. We thank Professor Takehiro Inohara of the Tokyo tech for providing helpful suggestions about GMCR and PMA. There are no conflicts of interest to declare associated with this manuscript or study. The datasets generated and analyzed during the current study are available from the corresponding author upon reasonable request.

References

1. Borda, J.-C.: Memorandum on the ballot box elections (in French). History of the Royal Academy of Sciences, Academy of Sciences (1781)
2. Sakai, T.: Invitation to Social Choice Theory (in Japanese). Nihon Hyoron Sya, Tokyo (2013)
3. Young, H.P.: Condorcet's theory of voting. Am. Polit. Sci. Rev. **82**(4), 1231–1244 (1988). https://doi.org/10.2307/1961757
4. Arrow, K.J.: Social Choice and Individual Values. Wiley, New York (1951)
5. Bentham, J.: An Introduction to the Principles of Morals and Legislation (Collected Works of Jeremy Bentham). Clarendon Press, Oxford (1996)
6. Rawls, J.: A Theory of Justice. Harvard University Press, Cambridge (1971)
7. Habermas, J.: Theory of communicative action. Band 2: On the critique of functionalist reason (in German). Suhrkamp Verlag, Frankfurt am Main (1981)
8. Fishkin, J.S.: When the People Speak: Deliberative Democracy and Public Consultation. Oxford University Press, Oxford (2009)
9. Gutmann, A., Thompson, D.: The Spirit of Compromise. Princeton University Press, Princeton (2012)
10. Ministry of Education, Culture, Sports, Science and Technology: Roundtable on Education Policy Formation Based on "Deliberative Discussion" (in Japanese). https://www.mext.go.jp/b_menu/shingi/chousa/shougai/022/index.htm. Accessed 29 Sept 2022
11. Cabinet Office: "New Public" Promotion Council (in Japanese). https://www5.cao.go.jp/npc/suishin.html. Accessed 29 Sept 2022
12. Thibaut, J.W., Walker, L.: Procedural Justice: A Psychological Analysis. Lawrence Erlbaum Associates, Hillsdale (1975)
13. Leventhal, G.S.: What should be done with equity theory? New approaches to the study of fairness in social relationships. In: Gergen, K.J., Greenberg, M.S., Willis, R.H. (eds.) Social Exchange. Springer, Boston (1980). https://doi.org/10.1007/978-1-4613-3087-5_2
14. Graeber, D.: There never was a west - or, democracy emerges from the spaces in between. In: Possibilities: Essays on Hierarchy, Rebellion, and Desire. AK Press, Oakland (2007)

15. Pickard, V.W.: Assessing the radical democracy of indymedia, discursive, technical, and institutional constructions. Crit. Stud. Media Commun. **23**(1), 19–38 (2006). https://doi.org/10.1080/07393180600570691
16. Susskind, L.E., Cruikshank, J.L.: Breaking Robert's Rules: The New Way to Run Your Meeting, Build Consensus. And Get Results. Oxford University Press, Oxford (2006)
17. Barandiaran, X.E., Calleja-Lopez, A., Monterde, A.: Decidim: political and technopolitical networks for participatory democracy. Decidim's project white paper, Barcelona (2018)
18. Hsiao, Y., Lin, S., Tang, A., Narayanan, D., Sarahe, C.: vTaiwan: An Empirical Study of Open Consultation Process in Taiwan. SocArXiv, Center for Open Science, Charlottesville (2018). https://doi.org/10.31235/osf.io/xyhft
19. Loomio: Loomio | decision-making for collaborative organizations. https://www.loomio.com/. Accessed 29 Sept 2022
20. Liquitous: Products aimed at democratic DX. https://liquitous.com/product/liqlid. Accessed 29 Sept 2022
21. Ito, T., Suzuki, S., Yamaguchi, N., Nishida, T., Hiraishi, K., Yoshino, K.: D-agree: crowd discussion support system based on automated facilitation agent. In: Proceedings of the AAAI Conference on Artificial Intelligence, vol. 34, no. 09, pp. 13614–13615. AAAI Press, Cambridge (2020). https://doi.org/10.1609/aaai.v34i09.7094
22. Ali, S., Xu, H., Xu, P., Ahmed, W.: Evolutional attitude based on option prioritization for conflict analysis of urban transport planning in Pakistan. J. Syst. Sci. Syst. Eng. **28**(3), 356–381 (2019). https://doi.org/10.1007/s11518-019-5413-0
23. Hipel, K.W., Fang, L., Kilgour, D.M.: The graph model for conflict resolution: reflections on three decades of development. Group Decis. Negot. **29**(1), 11–60 (2019). https://doi.org/10.1007/s10726-019-09648-z
24. Pourvaziri, M., Mahmoudkelayeh, S., Yousefi, S.: Proposing a genetic algorithm-based graph model for conflict resolution approach to optimize solutions in environmental conflicts. In: Proceedings of the 9th International Conference on Water Resources and Environment Research (2022). https://www.researchgate.net/publication/360463002. Accessed 27 Oct 2022
25. Xu, H., Hipel, K.W., Kilgour, D.M., Fang, L.: Conflict Resolution Using the Graph Model: Strategic Interactions in Competition and Cooperation. Studies in Systems, Decision and Control, vol. 153, Springer, New York (2018). https://doi.org/10.1007/978-3-319-77670-5
26. Inohara, T.: Graph model for conflict resolution - GMCR: the graph model for conflict resolution. In: Operations Research as a Management Science Research, vol. 58, no. 4, pp. 204–211. The Operations Research Society of Japan, Tokyo (2013). (in Japanese)
27. Xu, H., Addae, B. A., Wei, R.: Conflict resolution under power asymmetry in the graph model. In: Proceedings of the 20th International Conference on Group Decision and Negotiation, vol. 7, pp. 42-1 (2020)
28. Shahbaznezhadfard, M., Yousefi, S., Hipel, K.W., Hegazy, T.: Dynamic-based graph model for conflict resolution: systems thinking adaptation to solve real-world conflicts. In: Proceedings of the 20th International Conference on Group Decision and Negotiation (2020). https://www.researchgate.net/publication/361017860. Accessed 27 Oct 2022
29. Yamazaki, A., Inohara, T., Nakano, B.: The relationship between voter permissible range and the core of the simple game. J. Oper. Res. Soc. Jpn **42**(3), 286–301 (1999). https://doi.org/10.15807/jorsj.42.286. (in Japanese)
30. Yamazaki, A., Inohara, T., Nakano, B.: New interpretation of the core of simple games in terms of voters' permission. Appl. Math. Comput. **108**(2–3), 115–127 (2000). https://doi.org/10.1016/S0096-3003(99)00008-9
31. Corbyn, J.: Digital Democracy Manifesto. https://www.jeremyforlabour.com/digital_democracy_manifesto/. Accessed 29 Sept 2022

32. Schneider, N.: Everything for Everyone: The Radical Tradition that Is Shaping the Next Economy. Bold Type Books, New York (2018)
33. Asa, Y., Kato, T., Mine, R.: Composite Consensus-Building Process: Permissible Meeting Analysis and Compromise Choice Exploration. arXiv:2211.08593 (2022). https://doi.org/10.48550/arXiv.2211.08593
34. Astrachan, O.: Bubble sort: an archaeological algorithmic analysis. ACM SIGCSE Bull. **35**(1), 1–5 (2003). https://doi.org/10.1145/792548.611918
35. Saari, D.G., Merlin, V.R.: A geometric examination of Kemeny's rule. Soc. Choice Welfare **17**(3), 403–438 (2000). https://doi.org/10.1007/s003550050171

Equilibrium and Efficiency in Conflict Analysis Incorporating Permissibility

Yukiko Kato(✉)

Lynx Research Llc, Tokyo, Japan
kato.y.bj@m.titech.ac.jp

Abstract. In this study, we aim to generalize the Nash equilibrium and efficiency established in conflict resolutions among decision-makers with permissibility in their preferences for possible outcomes on the framework of GMCR(Graph Model for Conflict Resolution). Obtaining sufficient information on preferences, especially in emergent crises, can often be daunting in real-world conflicts. However, identifying "unacceptable situations" is comparatively less challenging. Our proposed approach dichotomizes preferences into binary categories of "permissible" and "impermissible," exhibiting a particular aptness for decision-making in situations with limited or focused information that seek to prevent severe crises, particularly during the emergent phase or convergence point of conflicts. We provide propositions on the equilibrium and efficiency of permissibility analysis, introducing a novel approach using coarse decision theory. Overall, our study contributes significantly to improving the convenience and effectiveness of real-world conflict analysis.

Keywords: GMCR · permissible range · coarse decision theory

1 Introduction

Real-world decision-making often requires quick first-order decisions to prevent worst-case scenarios, even in the absence of sufficient information for a detailed analysis. We aim to develop a decision-making approach based on analyzing coarse information. To achieve this, we have introduced several new concepts. Firstly, we propose a method to describe states where unknown factors other than the primary decision maker (DM) impact the DM's state transitions [1, 2]. Secondly, we introduced a new state recognition concept that expands the DM's controllable choices beyond the binary values of true (T) or false (F) to include both (B) and none (N), thus accommodating contradictions [3]. Finally, we presented a concept for incorporating permissible ranges (PR) in the DM's preferences [4]. This paper explores a novel analytical approach that employs Graph Model for Conflict Resolution (GMCR) [5,6] in situations where there is insufficient information available regarding DMs' preferences. Building on the

Y. Maemura et al. (Eds.): GDN 2023, LNBIP 478, pp. 113–129, 2023.
https://doi.org/10.1007/978-3-031-33780-2_8

foundational concept of PR proposed in our previous study [4], this research presents several propositions to extend and generalize the approach.

To address preference uncertainty, GMCR has developed various approaches based on pairwise relationships of states, such as unknown [7–10], fuzzy [11–14], grey [15–18], and probabilistic [19,20] methods. The use of matrix representation facilitates more intricate categorization calculations and effectively tackles these uncertainties. Various other approaches have also been examined to manage uncertain preferences, including setting a permissible range for alternatives based on the committee framework in the context of simple game [21–25].

Nonetheless, due to their unique nature, there are inherent limitations in dealing with severe crises. These crises are either unprecedented or infrequent, resulting in restricted access to the information required for analysis. Additionally, the severity of the crisis renders empirical testing of the model implausible. For instance, while retrospective analysis of the simultaneous terrorist attacks in 2001 is feasible, evaluating experimentally the conditions under which the events occurred is impractical. Assuming complete knowledge of environmental dynamics, the optimal response to risk can be achieved through dynamic programming based on state transitions and the utility derived from those transitions in a given scenario. However, in cases where the information available is limited and the worst-case scenario is catastrophic, the selection of an appropriate model and the information partitioning utilized in the model must be carefully considered, given the constraints.

This study is grounded on the premise that adopting a coarse framework is rational and practical for decision-making in situations where information is scarce, and the aim is to avoid worst-case scenarios. Concerning the resolution of severe conflicts to prevent worst-case scenarios, the current approaches, for addressing uncertain preferences in GMCR [7–20], tend to augment the information categories required for managing uncertainty, which is antithetical to the objective of this study. Furthermore, the simple game-based approach [21–25] is a framework that is efficacious in situations originally intended for cooperative resolution and may not necessarily be applicable to analyzing non-cooperative, severe conflict scenarios. Within the framework of GMCR, this study presents a new and innovative approach to analysis by employing a smaller number of information categories and proposing several key propositions. These findings not only advance our understanding of conflict resolution study but also have the potential to establish valuable links with other theoretical perspectives, including coarse decision theory.

Section 2 begins with an exposition of the foundational principles that underpin the current research, focusing specifically on the core concepts of coarse information and decision-making systems. Subsequently, we review the framework for conflict analysis incorporating permissibility, which provides basic concepts. We scrutinize the relationship between the DMs' PRs, equilibrium, and efficiency, and then present generally valid propositions. In Sect. 3, the validity of the propositions is verified by applying them to the case of the Elmira environmental dispute: the case most frequently discussed in GMCR studies.

2 Underlying Concepts and Methods

2.1 Coarse Decision Theory

There is a great deal of insight to be gained from literature in the fields of economics, finance, and psychology about the models and information partitioning that DMs adopt and their validity [26–34]. Among them, *rational inattention* of DMs under limited information processing capacity, proposed by Sims [26] is remarkable. A priori, we know that it is impossible to solve the problem of temporal imprecision when considering the common knowledge that is the premise of the state of the world in decision-making. In this sense, it can be said to be reasonable to use a coarser granularity [32]. The coarser the criteria (fewer categories for each criterion), the lower the decision-making cost, even though the DM has to use more criteria [33]. The maximum number of alternative distinctions that can be generated considering the number of categories for each criterion is equal to the product of the number of categories for the criterion deployed. Theoretically, it can be said that there is a trade-off between categories and criteria when considering an efficient decision-making function with a limited amount of information. Obviously, in the analysis aimed at a solution that avoids the worst-case scenario, which is the subject of this study, a more reasonable solution can be obtained by reducing the number of criteria.

2.2 Graph Model for Conflict Resolution (GMCR)

GMCR is a framework consisting of four tuples: $(N, S, (A_i)_{i \in N}, (\succsim_i)_{i \in N})$ [5,6]. N is the set of all DMs, S denotes the set of all feasible states. (S, A_i) constitutes DM i', s graph G_i, where S is the set of all vertices and $A_i \subset S \times S$ is the set of all oriented arcs. (S, A_i) has no loops; $(s, s) \in A$ for each $s \in S$. The preferences of each DM are presented as $(\succsim_i)$, where the set of all DMs $N : |N| \geq 2$, set of all states $S : |S| \geq 2$, and preference of DM i satisfy reflectiveness, completeness, and transitivity. $s \succsim_i s'$: s is equally or more preferred to s' by DM i; $s \succ_i s'$: s is strictly preferred to s' by DM i; $s \sim_i s'$: s is equally preferred to s' by DM i. We assume that a rational DM desires the situation to change to a more favorable state and attempts to transition to the preferred state by repeating unilateral moves, which the DM exercises control over. For $i \in N$ and $s \in S$, we define DM i's reachable list from state s as the set $\{s' \in S \mid (s, s') \in A_i\}$, denoted by $R_i(s)$. $R_i(s)$ is the set of all the states in which DM i can move from s to s' in a single step. A unilateral improvement of DM i from state s is defined as an element of the reachable list of DM i from s (i.e., $s' \in R_i(s)$), where i strictly prefers state s' ($s' \succ_i s$). Therefore, the set of the unilateral improvement lists of DM i from state s is described as $\{s' \in R_i(s) | s' \succ_i s\}$ and denoted by $R_i^+(s)$. $\phi_i^+(s)$ denotes the set of all states that are more preferential for DM i to s described as $\{s' \in S \mid s' \succ_i s\}$, and $\phi_i^{\simeq}(s)$ denotes the set of all states that are at most equally preferential to state s, described as $\{s' \in S \mid s \succsim_i s'\}$. Moreover, $R_{N-\{i\}}(s)$ is defined as the set of all states that can be achieved by the sequences of unilateral moves of DMs other than DM i. Similarly, $R_{N-\{i\}}^+(s)$ is defined as the set of all states that can be achieved by the sequences of unilateral improvements of DMs other than DM i.

On the basis of the DMs' state transitions, we can obtain standard stability concepts : Nash stability (**Nash**) [35,36], general meta-rationality (**GMR**) [37], symmetric meta-rationality (**SMR**) [37], and sequential stability (**SEQ**) [38,39].

2.3 GMCR Incorporating Permissible Range (GMCR-PR)

Using the elements of GMCR presented in the previous subsection, we now define GMCR-PR.

Based on the properties of preferences in GMCR mentioned in Sect. 2.2 (reflexivity, completeness, and transitivity), defining DM's permissibility as a weak order on the set S of all possible states, then a non-empty subset of the set $L(S)$ of all weak orders on S can represent the permissible preference of a DM. Specifically, for any $i \in N$, a subset P_i that satisfies $\emptyset \neq P_i \subseteq L(S)$ can be considered as the permissible states of DM i. We refer to P_i as the permission of DM i, which is defined as a subset of the set of all linear orderings that includes the DM's actual preferences. This definition of DM's permission is motivated by the recognition that the true preferences of DMs are not always accurately known in real-world group decision-making situations. While new definitions relating to improvement are introduced, the general definitions of GMCR such as DM i's reachable list $R_i(s)$ and $\phi_i^{\simeq}(s)$ provided in Sect. 2.2 remain unchanged.

Definition 1 (Permissible States (PS)). *For any $i \in N$, Permissible States (PS) of DM i is a non-empty subset of L, denoted by P_i. A list $(P_i)_{i \in N}$ for each $i \in N$ represents DMs' PS, denoted by P.*

By imposing a permissible threshold, the state set S can be partitioned into two subsets: those that are permissible for the DM and those that are not. This partition can be interpreted that $|P_i| = 1$.

Definition 2 (Permissible Range (PR)). *We denote DM i's PR by P_i^k, that is, in a conflict, DM i allows up to the kth most preferred state.*

GMCR-PR is represented by five tuples: DMs (N), a set of feasible states (S), a graph of DM i (A_i), the preferences of each DM i ($\succsim_i$), and a set of permissible preferences of DM i (P_i).

Definition 3 (GMCR-PR).

$$G = (N, S, (A_i)_{i \in N}, (\succsim_i)_{i \in N}, (P_i)_{i \in N}). \tag{1}$$

Example 1. Consider a conflict $G = (N, S, (A_i)_{i \in N}, (\succsim_i)_{i \in N}, (P_i)_{i \in N})$, where $N = \{a, b\}$, $S = \{1, 2, 3, 4\}$, $1 \succ_a 2 \succ_a 3 \succ_a 4$ and $4 \succ_b 3 \succ_b 2 \succ_b 1$. Suppose each DM's *PR* is P_a^2 and P_b^2 respectively, $P = \emptyset$.

Definition 4 (Reachable Lists in GMCR-PR). *DM i's* permissible reachable list *from $s \in S$ are subsets of S as follows:*

i. *DM i's reachable list from s to s′ by unilateral moves in GMCR-PR is defined as in GMCR,* $R_i(s) = \{s' \in S \mid (s, s') \in A_i\}$
ii. *DM i's Unilateral Improvement in GMCR-PR (PUI) is a transition from a state* $s \notin P_i$ *to a state* $s' \in P_i$ *and is defined as follows:*

$$PR_i(s) = \{s' \in R_i(s) \mid (s, s') \in A_i, s \notin P_i, s' \in P_i\}. \tag{2}$$

A list of PUI by a DM other than DM i is represented as $PR_{N-i}(s)$.
iii. *DM i's list of states regarding s and s′ being equally or less preferred is defined as in GMCR,* $\phi_i^{\precsim}(s) = \{s' \in S \mid s \succsim_i s'\}$.

When DM i has no PUI from state s, there are no further state transitions exist, thereby establishing stability.

Definition 5 (PNash). *For* $i \in N$*, state* $s \in S$ *is PNash stable for DM i, denoted by* $s \in S_i^{PNash}$*, if and only if*

$$PR_i(s) = \emptyset. \tag{3}$$

State s is PGMR stable for DM i when any PUI from state s of DM i may cause a state equal or less preferred state than s in the responses of the other DMs.

Definition 6 (PGMR). *For* $i \in N$*, state* $s \in S$ *is PGMR stable for DM i, denoted by* $s \in S_i^{PGMR}$*, if and only if*

$$\forall s' \in PR_i(s), R_{N-\{i\}}(s') \cap \phi_i^{\precsim}(s) \neq \emptyset. \tag{4}$$

When a state occurs where for any of DM i's PUI, another DM's countermove would result in a state equal or less favorable than s. Furthermore, regardless DM i's subsequent countermove, a state more favorable than s cannot occur, thereby establishing stability.

Definition 7 (PSMR). *For* $i \in N$*, state* $s \in S$ *is PSMR stable for DM i, denoted by* $s \in S_i^{PSMR}$*, if and only if*

$$\forall s' \in PR_i(s), \exists s'' \in R_{N-\{i\}}(s') \cap \phi_i^{\precsim}(s),\ R_i(s'') \subseteq \phi_i^{\precsim}(s). \tag{5}$$

When DM i has at least one PUI from state s, but states resulting from PUI of other DMs' responses from s' cause a state to be equal to or less preferable than s for DM i, then s is PSEQ stable for DM i.

Definition 8 (PSEQ). *For* $i \in N$*, state* $s \in S$ *is PSEQ stable for DM i, denoted by* $s \in S_i^{PSEQ}$*, if and only if*

$$\forall s' \in PR_i(s),\ PR_{N-\{i\}}(s') \cap \phi_i^{\precsim}(s) \neq \emptyset. \tag{6}$$

The chicken game in GMCR-PR can be represented as follows.

Example 2 (Chicken Game).
$(N, S, (A_i)_{i \in N}, (\succsim_i)_{i \in N}), N = \{1, 2\}, S = \{s_1, s_2, s_3, s_4\}$
$A_1 = \{(s_1, s_3), (s_3, s_1), (s_2, s_4), (s_4, s_2)\}$,
$A_2 = \{(s_1, s_2), (s_2, s_1), (s_3, s_4), (s_4, s_3)\}$,
DM_1's preference order $\succsim_1$: $s_3 \succ s_1 \succ s_2 \succ s_4$,
DM_2's preference order $\succsim_2$: $s_2 \succ s_1 \succ s_3 \succ s_4$.

Let us assume that both DMs have permissibility up to the states of the second preference order P^2, in stead of the linear order provided in the original game: $P_1 = \{s_1, s_3\}$, $P_2 = \{s_1, s_2\}$. Then, we have unilateral improvements for each DM as follows; $PR_1(s_1) = \emptyset, PR_1(s_2) = \emptyset, PR_1(s_3) = \emptyset, PR_1(s_4) = \emptyset$, $PR_2(s_1) = \emptyset, PR_2(s_2) = \emptyset, PR_2(s_3) = \emptyset, PR_2(s_4) = \emptyset$. Hence, Nash equilibrium is established in all states when P^2 is employed for each DM in the chicken game. The Table 1 summarizes the permissibility, reachability, PUI, and Nash equilibrium.

Table 1. Chicken Game in GMCR-PR - P_1^2, P_2^2

State		1	2	3	4
Permissibility	DM1	1	0	1	0
	DM2	1	1	0	0
$R_i(s)$	DM1	3	4	1	2
	DM2	2	1	4	3
$\phi_i^+(s)$	DM1		1,3		1,3
	DM2			1,2	1,2
PR_i	DM1				
	DM2				
Nash			E	E	
PNash		E	E	E	E

In our previous study [4], we examined the equilibrium and efficiency of 21 sets of 2×2 games classified as Class III in Rapoport and Guyer's taxonomy [40], in which neither DM has a dominant strategy for all combinations of the four permissible levels. Consequently, any conflict in the category was concluded to be resolved when both DMs set their threshold as P^3: "accept all states except the least favorable one." The following section develops the discussions on permissibility and equilibrium/efficiency based on these results and present propositions. Incorporating the concept of "permissibility" in the derivation of propositions concerning conflict resolution would enrich the spectrum of recommendations for resolving conflicts.

3 Nash Stability and Efficiency for Conflicts with Permissible Range

We present propositions for Nash stability and efficiency in conflict analysis incorporating PR. These propositions were prepared separately for cases with at least one state commonly permissible to all DMs in Subsect. 3.1, and cases without such a state in Subsect. 3.2.

Consider a conflict presented in GMCR-PR: $(N, S, (A_i)_{i\in N}, (\succsim_i)_{i\in N}, (P_i))$. Here, for $i \in N$, P_i denotes the set of all permissible states for DM i; Therefore, if $s \in S$ is permissible for DM i, then it is denoted by $s \in P_i$; otherwise, $s \notin P_i$.

3.1 Case with $\cap_{i\in N} P_i \neq \emptyset$

First, we consider the case with $\cap_{i\in N} P_i \neq \emptyset$; that is, there exists at least one state that is commonly permissible for all DMs. We have the following propositions:

Nash Stability

Proposition 1. State $s \in \cap_{i\in N} P_i$ is Nash equilibrium.

Proof. For $i \in N$, we have $R_i^+(s) = \emptyset$, because for all $s' \in S$, $s \succsim_i s'$. □

Proposition 2. Consider state $s' \notin \cap_{i\in N} P_i$. For $j \in N$, if $s' \in P_j$, then s' is Nash stable for DM j.

Proof. s' is Nash stable for DM j, because for all $s'' \in S$, $s' \succsim_j s''$. □

Proposition 3. Consider state $'s \notin \cap_{i\in N} P_i$. For $k \in N$, if $s' \notin P_j$, then s' is Nash stable for DM k if $R_k(s') \cap P_k = \emptyset$, and not if $R_k(s') \cap P_k \neq \emptyset$.

Proof. s' is Nash stable for DM k if $R_k(s') \cap P_k = \emptyset$, because we have $R_k^+(s') = \emptyset$ from $s' \notin P_k$ and for all $s'' \in R_k(s')$, $s'' \notin P_k$. s' is not Nash stable for DM k if $R_k(s') \cap P_k \neq \emptyset$, because we have $R_k^+(s') \neq \emptyset$ from $s' \notin P_k$ and there exists $s'' \in P_k(s')$ such that $s'' \in P_k$, which implies $s'' \succ_k s'$. □

For special cases in which each DM's PR includes all states except the least preferable one, we have Corollary 1 of Proposition 3.

Corollary 1 (Corollary of Proposition 3).
Consider cases that $P_i = S\backslash\{\min \succsim_i\}$ *for* $i \in N$, *where* $\min \succsim_i$ *denotes DM* i*'s least preferred state.* $s' = \min \succsim_i$ *is Nash stable for DM* i *if* $R_i(s') = \emptyset$, *and not if* $R_i(s') \neq \emptyset$.

Proof. If $R_i(s') = \emptyset$, then we always have $R_i^+(s') = \emptyset$, which means that s' is Nash stable for DM i. If $R_i(s') \neq \emptyset$, then we have that $R_i(s') \cap P_i \neq \emptyset$, because $R_i(s') \subseteq S\backslash\{s'\} = S\backslash\{\min \succsim_i\} = P_i$. Using the result of Proposition 3, we have that s' is not Nash stable for DM i. □

Efficiency. The following are propositions on the efficiency of states under the condition of $\cap_{i \in N} P_i \neq \emptyset$

Proposition 4. State $s \in \cap_{i \in N} P_i$ is weakly and strongly efficient.

Proof. In this case, for all $i \in N$ and all $s' \in S$, $s \succsim_i s'$. Therefore, $s' \succ_i s$ cannot be satisfied for any $i \in N$ and any $s' \in S$, which implies that s is weakly and strongly efficient. □

Proposition 5. Consider state $s' \notin \cap_{i \in N} P_i$. For $j \in N$, if $s' \in P_j$ (which implies that $s' \notin P_k$ for some $k \in N$), then s' is weakly efficient and not strongly efficient.

Proof. In this case, for all $i \in N$, $s \succsim_i s'$ and $s \succ_k s'$, because $s \in \cap_{i \in N} P_i$ and $s' \notin P_k$. This implies that s' is not strongly efficient. No $s'' \in S$ exists such that $s'' \succ s'$ for all $i \in N$, because $s' \in P_j$. This implies that s' is weakly efficient. □

Proposition 6. Consider state $s' \notin \cap_{i \in N} P_i$. If $s' \notin P_i$ for all $i \in N$, then s' is neither weakly nor strongly efficient.

Proof. In this case, for all $i \in N$, $s \succ_i s'$, because for all $i \in N$, and $s \in P_i$ and for all $i \in N$, $s' \notin P_i$. □

Worst Case Efficiency. For situations in which each DM's PR includes all cases except the least preferable one, we have Corollary 2 of Propositions 5 and 6.

Corollary 2 (Corollary of Propositions 5 and 6).
Consider the cases in which $P_i = S \backslash \{\min \succsim_i\}$ *for all* $i \in N$, *where* $\min \succsim_i$ *denotes DM* i*'s least preferred state.* $\min \succsim_i$ *is weakly efficient and not strongly efficient if* $\min \succsim_i \neq \min \succsim_j$ *for some* i *and* $j \in N$. $\min \succsim_i$ *is neither weakly nor strongly efficient if* $\min \succsim_i = \min \succsim_j$ *for all* i *and* $j \in N$.

Proof. In the case in which $\min \succsim_i \neq \min \succsim_j$ for some i and $j \in N$, $s' = \min \succsim_i \notin P_i$ and $s' \in P_j$. Then, by applying Proposition 5, we have that $\min \succsim_i$ is weakly efficient and not strongly efficient.

In the case in which $\min \succsim_i = \min \succsim_j$ for all i and $j \in N$, $s' = \min \succsim_i \notin P_i$ for all $i \in N$. Then, by applying Proposition 6, we have that $\min \succsim_i$ is neither weakly nor strongly efficient. □

3.2 Case with $\cap_{i \in N} P_i = \emptyset$

Next, we consider the case with $\cap_{i \in N} P_i = \emptyset$, that is, there is no state that is commonly permissible for all DMs exists. We have the following propositions:

Nash Stability

Proposition 7. Consider state $s' \notin \cap_{i \in N} P_i$. For $j \in N$, if $s' \in P_j$, then s' is Nash stable for DM j.

Proof. s' is Nash stable for DM j because for all $s'' \in S$, $s' \succsim_j s''$. □

Proposition 8. Consider state $s' \notin \cap_{i \in N} P_i$. For $k \in N$, if $s' \notin P_j$, then s' is Nash stable for DM k if $R_k(s') \cap P_k = \emptyset$, and not if $R_k(s') \cap P_k \neq \emptyset$.

Proof. s' is Nash stable for DM k if $R_k(s') \cap P_k = \emptyset$, because we have $R_k^+(s') = \emptyset$ from $s' \notin P_k$ and for all $s'' \in R_k(s')$, $s'' \notin P_k$. s' is not Nash stable for DM k if $R_k(s') \cap P_k \neq \emptyset$, because we have $R_k^+(s') \neq \emptyset$ from $s' \notin P_k$ and there exists $s'' \in P_k(s')$ such that $s'' \in P_k$, which implies $s'' \succ_k s'$. □

For situations in which each DM's PR includes all states except the least preferable one, we have Corollary 3 of Proposition 8.

Corollary 3 (Corollary of Proposition 8).
Consider the cases in which $P_i = S \backslash \{\min \succsim_i\}$ *for all* $i \in N$*, where* $\min \succsim_i$ *denotes DM* i*'s least preferred state.* $s' = \min \succsim_i$ *is Nash stable for DM* i *if* $R_i(s') = \emptyset$*, and not Nash stable if* $R_i(s') \neq \emptyset$.

Proof. If $R_i(s') = \emptyset$, then we always have $R_i^+(s') = \emptyset$, which means that s' is Nash stable for DM i. If $R_i(s') \neq \emptyset$, then we have $R_i(s') \cap P_i \neq \emptyset$, because $R_i(s') \subseteq S \backslash \{\min \succsim_i\} = P_i$. Using the result of Proposition 8, we have that s' is not Nash stable for DM i. □

Efficiency. The following are propositions on the efficiency of states under the condition of $\cap_{i \in N} P_i = \emptyset$

Proposition 9. Consider state $s' \notin \cap_{i \in N} P_i$. For $j \in N$, if $s' \in P_j$ (which implies that $s' \notin P_k$ for some $k \in N$), then s' is weakly efficient.

Proof. No $s'' \in S$ exists such that $s'' \succ_i s'$ for all $i \in N$, because $s' \in P_j$. Thus, s' is weakly efficient. □

Proposition 10. Consider state $s' \notin \cap_{i \in N} P_i$. Assume that $N = \{j, k\}$, that is $|N| = 2$. Then, for $j \in N$, if $s' \in P_j$ (which implies that $s' \notin P_k$ for the other $k \in N$), then s' is strongly efficient.

Proof. Assume that there exists $s'' \in S$ such that $s'' \succsim_j s'$ and $s'' \succsim_k s'$, and that $s'' \succ_j s'$ or $s'' \succ_k s'$. Because $s' \in P_j$, it is impossible that $s'' \succ_j s'$. This implies that $s'' \succ_k s'$. Then, we must have that $s'' \in P_j$ and $s'' \in P_k$, which contradicts the condition that $\cap_{i \in N} P_i = \emptyset$. Therefore, s' is strongly efficient. □

With respect to the strong efficiency of state s' under the conditions of $\cap_{i \in N} P_i = \emptyset$, $s' \in P_j$ for some $j \in N$, $s' \notin P_k$ for some $k \in N$, and $|N| \geq 3$, see the following example. We see that s' may be strongly efficient depending on $(P_i)_{i \in N}$ in the following examples.

Example 3.

Case 1: Let $N = \{1, 2, 3\}$, $S = \{s_1, s_2, s_3\}$, and $P_1 = \{s_1, s_2\}$; $P_2 = \{s_2\}$; $P_3 = \{s_3, s_1\}$. In this case, $\cap_{i \in N} P_i = \emptyset$, and $s_1 \in P_1$; $s_1 \notin P_2$; $s_1 \notin P_3$. We see that s_1 is strongly efficient, because $s_1 \succ_3 s_2$ and $s_1 \succ_1 s_3$.
Case 2: Let $N = \{1, 2, 3\}$, $S = \{s_1, s_2, s_3\}$, and $P_1 = \{s_1, s_2\}$; $P_2 = \{s_2\}$; $P_3 = \{s_3\}$. In this case, $\cap_{i \in N} P_i = \emptyset$, and $s_1 \in P_1$; $s_1 \notin P_2$; $s_1 \notin P_3$. We see that s_1 is not strongly efficient because $s_2 \succsim_1 s_1$; $s_2 \succ_2 s_1$; $s_2 \succsim_3 s_1$.

Proposition 11. Consider state $s' \notin \cap_{i \in N} P_i$. If $s' \notin P_i$ for all $i \in N$, then s' is weakly efficient and not strongly efficient.

Proof. Assume that there exists $s'' \in S$ such that for all $i \in N$, $s'' \succ_i s'$. Then, we must have that for all $i \in N$, $s'' \in P_i$, which contradicts the condition that $\cap_{i \in N} P_i = \emptyset$. Thus, s' is weakly efficient. Because we assume that $P_j \neq \emptyset$ for all $j \in N$, we can take $s'' \in P_j$. Then, it is satisfied that $s'' \succ_j s'$ and $s'' \succsim_i s'$ for all $i \in N$, because $s' \notin P_i$ for all $i \in N$. Therefore, s' is not strongly efficient. □

Worst Case Efficiency

Corollary 4 (Corollary of Proposition 9 and Proposition 10). *Consider the cases that $P_i = S \backslash \{\min \succsim_i\}$ for all $i \in N$, in which $\min \succsim_i$ denotes DM i's least preferred state. Then, we have that $\min \succsim_i$ is weakly efficient. We also have that $\min \succsim_i$ is strongly efficient if $N = \{1, 2\}$.*

Proof. Under the conditions of $\cap_{i \in N} P_i = \emptyset$ and $P_i = S \backslash \{\min \succsim_i\}$ for all $i \in N$, we have that $S = \{\min \succsim_i \mid i \in N\}$, because otherwise, $x \in S \backslash \{\min \succsim_i \mid i \in N\}$ satisfies that $x \in \cap_{i \in N} P_i$, which contradicts the condition that $\cap_{i \in N} P_i = \emptyset$. Then, $S = \{\min \succsim_i \mid i \in N\}$ implies the results using Propositions 9 and 10. □

For strong efficiency in cases with $P_i = S \backslash \{\min \succsim_i\}$ for all $i \in N$ and $|N| \geq 3$, we have the following proposition:

Proposition 12. Consider cases that $P_i = S \backslash \{\min \succsim_i\}$ for all $i \in N$, where $\min \succsim_i$ denotes the DM i's least preferred state. Then, we have that $\min \succsim_i$ is strongly efficient, if $|N| \geq 3$.

Proof. Under the conditions of $\cap_{i \in N} P_i = \emptyset$ and $P_i = S \backslash \{\min \succsim_i\}$ for all $i \in N$, we have that $S = \{\min \succsim_i \mid i \in N\}$, because otherwise, $x \in S \backslash \{\min \succsim_i \mid i \in N\}$ satisfies $x \in \cap_{i \in N} P_i$, which contradicts the condition that $\cap_{i \in N} P_i = \emptyset$.

For all $s'' \in S = \{\min \succsim_i \mid i \in N\}$, there exists $i \in N$ such that $s'' = \min \succsim_i$, which implies that $s' \succ_i s''$. □

Table 2 summarizes the results for general cases in Subsects. 3.1 and 3.2, and Table 3 shows the results for the cases with $P_i = S \backslash \{\min \succsim_i\}$ for all $i \in N$ given by the corollaries in Subsects. 3.1 and 3.2.

Table 2. Interrelationships between Nash Stability and Efficiencies

	$s \in S$: $s \in \cap_{i \in N} P_i$	$s' \in S$: $s' \in P_j$ and $s' \notin P_k$	$s' \in S$: $\forall k \in N, s' \notin P_k$
If $\cap_{i \in N} P_i \neq \emptyset$:	Nash for all $i \in N$ (Proposition 1)	Nash for j (Proposition 2)	—
		Nash for k depending on $R_k(s')$ and P_k (Proposition 3)	
	w.eff. (Proposition 4)	w.eff. (Proposition 5)	NOT w.eff. (Proposition 6)
	s.eff. (Proposition 4)	NOT s.eff. (Proposition 5)	NOT s.eff. (Proposition 6)
If $\cap_{i \in N} P_i = \emptyset$:	—	Nash for j (Proposition 7)	—
		Nash for k depending on $R_k(s')$ and P_k (Proposition 8)	
	—	w.eff. (Proposition 9)	w.eff. (Proposition 11)
	—	s.eff. if $\lvert N\rvert = 2$ (Proposition 10); dep.on $(P_i)_{i \in N}$ if $\lvert N\rvert \geq 3$ (Ex. 3)	NOT s.eff. (Proposition 11)

Table 3. Nash stability and efficiencies of min $\succsim_i$ under the condition of $P_i = S \backslash \{\min \succsim_i\}$ for all $i \in N$

	$\exists i, j \in N$, min $\succsim_i \neq$ min $\succsim_j$	$\forall i, j \in N$, min $\succsim_i =$ min $\succsim_j$
If $\cap_{i \in N} P_i \neq \emptyset$:	Nash for i depending on $R_i(s')$ (Corollary 1)	
	w.eff. (Corollary 2)	NOT w.eff. (Corollary 2)
	NOT s.eff. (Corollary 2)	NOT s.eff. (Corollary 2)
If $\cap_{i \in N} P_i = \emptyset$:	Nash for i depending on $R_i(s')$ (Corollary 3)	
	w.eff. (Corollary 4)	—
	s.eff. (Corollary 4, Proposition 12)	—

4 Verification of Propositions in Application Cases

In this section, the propositions presented in Sect. 3 are verified by applying them to the Elmira conflict, a representative case of GMCR analysis [41,42].

Elmira Conflict. The Elmira conflict is an environmental contamination dispute in Ontario, Canada, upon which numerous studies have been conducted using GMCR. Three DMs are involved in the conflict: the Ministry of Environment (**M**), Uniroyal (**U**), and the local government (**L**). **M** discovered contamination and issued a control order to **U** that included a decontamination operation to be conducted by **U**. They desire to exercise their authority efficiently. **U** operates questionable chemical plants, and intends to exercise its right to object, aiming to lift or relax the control order. **L** represents diverse interest groups, and intends to protect the residents and the local industrial base. Table 4 summarizes all feasible states based on the DMs' options, while (Fig. 1) displays the corresponding conflict graph. In addition, the preference orders of the three DMs are given as follows: **M** : $s_7 \succ s_3 \succ s_4 \succ s_8 \succ s_5 \succ s_1 \succ s_2 \succ s_6 \succ s_9$;

$\mathbf{U}: s_1 \succ s_4 \succ s_8 \succ s_5 \succ s_9 \succ s_3 \succ s_7 \succ s_2 \succ s_6$; $\mathbf{L}: s_7 \succ s_3 \succ s_5 \succ s_1 \succ s_8 \succ s_6 \succ s_4 \succ s_2 \succ s_9$.

Table 4. Elmira Conflict - Options and States

State		1	2	3	4	5	6	7	8	9
M	Modify	N	Y	N	Y	N	Y	N	Y	-
U	Delay	Y	Y	N	N	Y	Y	N	N	-
	Accept	N	N	Y	Y	N	N	Y	Y	-
	Abandon	N	N	N	N	N	N	N	N	Y
L	Insist	N	N	N	N	Y	Y	Y	Y	-

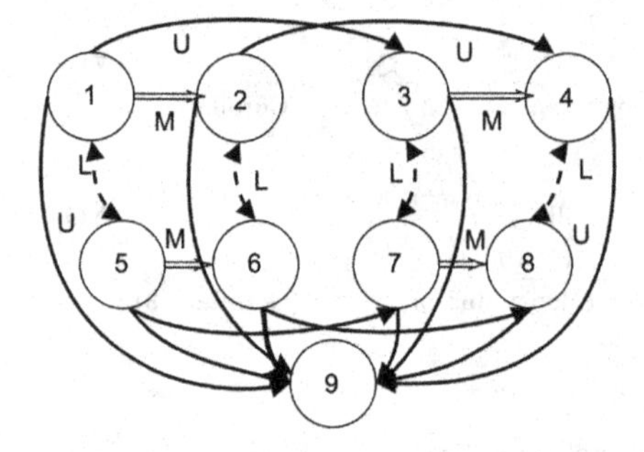

Fig. 1. Graph Model of Elmira conflict

Elmira Conflict Case-1: $\cap_{i \in N} P_i \neq \emptyset$. We verified the propositions presented in Sect. 3.1 for the case with $\cap_{i \in N} P_i \neq \emptyset$ by examining the stability analysis of the Elmira conflict case-P_M^2, P_U^7, and P_L^5. Table 5 summarizes the permissibility, reachability, and PNash.

In conflicts where at least one state is permissible to all DMs, we determined the following propositions: 1) Proposition 1 concerns the permissible states for all DMs; thus, Nash equilibria are established at s_3 and s_7. 2) Propositions 2 and 3 concern states other than those verified in 1) that are permissible for each DM, and lead to Nash stability in s_1, s_5, s_8, and s_9. From 1) and 2), we can conclude that PNash equilibria hold for s_1, s_3, s_5, s_7, s_8, and s_9. This verification result is consistent with the GMCR-PR stability analysis presented in Table 5. In addition, we observe that the weak and strong Pareto efficiency proposed in Proposition 4 is consistent with the original results in Table 5. Each item in the table indicates the following.

- Permissibility: Boolean value denoting permissibility for DM i.

Table 5. Verification of Propositions: Elmira Conflict - P_M^2, P_U^7, P_L^5

State		1	2	3	4	5	6	7	8	9
Preference order	M	7	3	4	8	5	1	2	6	9
	U	1	4	8	5	9	3	7	2	6
	L	7	3	5	1	8	6	4	2	9
Permissibility	M	0	0	1	0	0	0	1	0	0
	U	1	0	1	1	1	0	1	1	1
	L	1	0	1	0	1	0	1	1	0
$R_i(s)$	M	2		4		6		8		
	U	3,9	$\overline{4}, \overline{9}$	9	9	7,9	$\overline{8}, \overline{9}$	9	9	
	L	5	6	7	$\overline{8}$	1	2	3	4	
PNash Equilibrium		E		E		E		E	E	E
Proposition 1(Nash)				E				E		
Proposition 2(Nash)		U,L			U	U,L			U,L	U
Proposition 3(Nash)		M			M	M			M	M,L
Proposition 4(eff.)				✓				✓		
Proposition 5(eff.)					✓					✓
Proposition 6(eff.)			✓				✓			

- $R_i(s)$: States where DM i can unilaterally transition (UM) from each state. The numbers indicate the number of states. The overbar signifies that the transition is PUI.
- PNash Equilibrium: Nash holds for all $i \in N$
- Prop. 1–3(Nash): E denotes equilibrium, M, U, and L indicate DMs who reached stability according to the proposition.
- Prop. 4–6(eff.): Weak and strong efficiency holds for the state with the checkmark.

Elmira Conflict Case-2: $\bigcap_{i \in N} P_i = \emptyset$. We verified the propositions by setting up a PR case P_M^2, P_U^2, and P_L^2 in the Elmira conflict.

Table 6 presents the stability analysis when the PR of all DMs is set to P^2. It is presented as a conflict without a single state that is commonly permissible for all DMs. Table 7 presents the correspondence between the stability and propositions in the P_M^2, P_U^2, P_L^2 case.

In conflicts where no state is permissible to all DMs, we determined the following regarding the propositions: 1) Proposition 7 is about permissible states for DM j; thus, Nash stability holds at s_1 for **U**, s_3 for **M** and **L**, s_4 for **U**, and s_7 for **M** and **L**. 2) Propositions 8 concerns states other than those verified in 1) that are permissible for DM j, and this proposition leads to Nash stability for **M**, **U**, and **L**. From 1) and 2), we can conclude that the PNash equilibra hold in s_1, s_3, s_4, s_5, s_6, s_7, s_8 and s_9. This verification result is consistent with the

Table 6. Elmira Conflict - Stability Analysis: P_M^2, P_U^2, P_L^2

State	1	2	3	4	5	6	7	8	9
M	0	0	1	0	0	0	1	0	0
U	1	0	0	1	0	0	0	0	0
L	0	0	1	0	0	0	1	0	0
PNash	✓		✓	✓	✓	✓	✓	✓	✓
PGMR	✓	✓	✓	✓	✓	✓	✓	✓	✓
PSMR	✓	✓	✓	✓	✓	✓	✓	✓	✓
PSEQ	✓		✓	✓	✓	✓	✓	✓	✓
Pareto	✓		✓	✓			✓		

Table 7. Verification of Propositions: Elmira Conflict - P_M^2, P_U^2, P_L^2

State		1	2	3	4	5	6	7	8	9
Preference order	M	7	3	4	8	5	1	2	6	9
	U	1	4	8	5	9	3	7	2	6
	L	7	3	5	1	8	6	4	2	9
Permissibility	M	0	0	1	0	0	0	1	0	0
	U	1	0	0	1	0	0	0	0	0
	L	0	0	1	0	0	0	1	0	0
$R_i(s)$	M	2		4		6		8		
	U	3,9	$\overline{4}$,9	9	9	7,9	8,9	9	9	
	L	5	6	7	8	1	2	3	4	
PNash Equilibrium		E		E	E	E	E	E	E	E
Proposition 7(Nash)		U		M,L	U			M,L		
Proposition 8(Nash)		M,L	M,L	U	M,L	M,U,L	M,U,L	U	M,U,L	M,U,L
Proposition 9(eff.)		✓		✓	✓			✓		
Proposition 11(eff.)			✓			✓	✓		✓	✓

GMCR-PR stability analysis shown in Table 6. In addition, we seek to confirm that the weak and strong Pareto efficiencies provided in Propositions 9 and 11 are consistent with the original results in Table 6.

This section examined the propositions presented in Sect. 3 and verified it to be consistent with the results of the GMCR-PR stability analyses of the Elmira conflict.

5 Conclusion

This study discussed the analysis capability with coarse information by introducing the concept of PR to GMCR. PR is set by placing a threshold on the preference, and the DM's preference is processed as binary information. Moreover, because the GMCR framework is retained, the resolution can be changed depending on the granularity of the information available. Introducing the concept of PR allows for the analysis to reflect implicit assumptions that are not part of the fundamental framework; describing a situation in which even a DM seeking reasonable resolution endeavors to avoid prolongation or escalation to converge the conflict by adjusting its permissible level is possible.

This paper focused on equilibrium and efficiency in the two cases of the presence or absence of commonly permissible states for all DMs. Future research topics include more complex issues, such as those in which permissibility differs from the initial judgment because of the availability of information after the determination from the first analysis.

References

1. Kato, Y.: New reachability via the influence of external factors for conflict escalation and de-escalation. In: IEEE Systems, Man, and Cybernetics, Melbourne, pp. 813–819 (2021). https://doi.org/10.1109/SMC52423.2021.9659282
2. Kato, Y.: State transition for de-escalation in the graph model for conflict resolution framework. JSIAM Lett. **13**, 60–63 (2021). https://doi.org/10.14495/jsiaml.13.60
3. Kato, Y.: State definition for conflict analysis with four-valued logic. In: IEEE Systems, Man, and Cybernetics, Prague, pp. 3186–3191 (2022). https://doi.org/10.1109/SMC53654.2022.9945371
4. Kato, Y.: Binary processing of permissible range in graph model of conflict resolution. In: IEEE Systems, Man, and Cybernetics, Melbourne, pp. 685–690 (2021). https://doi.org/10.1109/SMC52423.2021.9658701
5. Fang, L., Hipel, K.W., Kilgour, D.M.: Interactive Decision Making: The Graph Model for Conflict Resolution. Wiley, New York (1993)
6. Kilgour, D.M., Hipel, K.W., Fang, L.: The graph model for conflicts. Automatica **23**(1), 41–55 (1987). https://doi.org/10.1016/0005-1098(87)90117-8
7. Li, K.W., Hipel, D.M., Kilgour, L.F.: Preference uncertainty in the graph model for conflict resolution. IEEE Trans. Syst. Man Cybern. Part A Syst. Hum. **34**(4), 507–519 (2004). https://doi.org/10.1109/TSMCA.2004.826282
8. Xu, H., Kilgour, D.M., Hipel, K.W.: Matrix representation of conflict resolution in multiple-decision-maker graph models with preference uncertainty. Gr. Decis. Negot. **20**(6), 755–779 (2011). https://doi.org/10.1007/s10726-010-9188-4
9. Al-Mutairi, M.S., Hipel, K.W., Kamel, M.S.: Fuzzy preferences in conflicts. J. Syst. Sci. Syst. Eng. **17**(3), 257–276 (2008). https://doi.org/10.1007/s11518-008-5088-4
10. Al-Mutairi, M.: Preference uncertainty and trust in decision making. Dissertation Abstracts International Section B Sciences and Engineering (2008)
11. Hipel, K.W., Kilgour, D.M., Bashar, M.A.: Fuzzy preferences in multiple participant decision making. Sci. Iran. **18**(3), 627–638 (2011). https://doi.org/10.1016/j.scient.2011.04.016
12. Bashar, M.A., Kilgour, D.M., Hipel, K.W.: Fuzzy preferences in the graph model for conflict resolution. IEEE Trans. Fuzzy Syst. **20**(4), 760–770 (2012). https://doi.org/10.1109/TFUZZ.2012.2183603
13. Bashar, M.A., Kilgour, D.M., Hipel, K.W.: Fuzzy option prioritization for the graph model for conflict resolution. Fuzzy Sets Syst. **246**, 34–48 (2014). https://doi.org/10.1016/j.fss.2014.02.011
14. Bashar, M.A., Obeidi, A., Kilgour, D.M., Hipel, K.W.: Modeling fuzzy and interval fuzzy preferences within a graph model framework. IEEE Trans. Fuzzy Syst. **24**(4), 765–778 (2016). https://doi.org/10.1109/TFUZZ.2015.2446536

15. Bashar, M.A., Hipel, K.W., Kilgour, D.M., Obeidi, A.: Interval fuzzy preferences in the graph model for conflict resolution. Fuzzy Optim. Decis. Making **17**(3), 287–315 (2017). https://doi.org/10.1007/s10700-017-9279-7
16. Kuang, H., Bashar, M.A., Hipel, K.W., Kilgour, D.M.: Grey-based preference in a graph model for conflict resolution with multiple decision makers. IEEE Trans. Syst. Man Cybern. Syst. **45**(9), 1254–1267 (2015). https://doi.org/10.1109/TSMC.2014.2387096
17. Kuang, H., Hipel, K.W., Kilgour, D.M., Bashar, M.A.: A case study of grey-based preference in a graph model for conflict resolution with two decision makers. In: IEEE Systems, Man, and Cybernetics, Manchester, UK, pp. 2037–2041 (2013). https://doi.org/10.1109/SMC.2013.349
18. Zhao, S., Xu, H.: Grey option prioritization for the graph model for conflict resolution. J. Grey Syst. **29**, 14 (2017)
19. Rêgo, L.C., dos Santos, A.M.: Upper and lower probabilistic preferences in the graph model for conflict resolution. Int. J. Approx. Reason. **98**, 96–111 (2018). https://doi.org/10.1016/j.ijar.2018.04.008
20. Rêgo, L.C., Vieira, G.I.A.: Probabilistic option prioritizing in the graph model for conflict resolution. Group Decis. Negot. **28**(6), 1149–1165 (2019). https://doi.org/10.1007/s10726-019-09635-4
21. Inohara, T.: Generalizations of the concept of core of simple games and their characterization in terms of permission of voters. Appl. Math. Comput. **132**(1), 47–62 (2002). https://doi.org/10.1016/S0096-3003(01)00175-8
22. Yamazaki, A., Inohara, T., Nakano, B.: Comparability of coalitions in committees with permission of voters by using desirability relation and hopefulness relation. Appl. Math. Comput. AMC **113**, 219–234 (2000). https://doi.org/10.1016/S0096-3003(99)00089-2
23. Yamazaki, A., Inohara, T., Nakano, B.: Symmetry of simple games and permission of voters. Appl. Math. Comput. AMC **114**, 315–327 (2000). https://doi.org/10.1016/S0096-3003(99)00126-5
24. Inohara, T.: Majority decision making and the graph model for conflict resolution. IEEE Syst. Man Cybern. (2011). https://doi.org/10.1109/ICSMC.2011.6084081
25. Inohara, T.: Consensus building and the graph model for conflict resolution. IEEE Syst. Man Cybern. (2010). https://doi.org/10.1109/ICSMC.2010.5641917
26. Sims, C.A.: Implications of rational inattention. J. Monet. Econ. **50**(3), 665–690 (2003). https://doi.org/10.1016/S0304-3932(03)00029-1
27. Pawlak, Z.: Rough sets. Int. J. Comput. Inform. Sci. **11**(5), 341–356 (1982). https://doi.org/10.1007/bf01001956
28. Ziarko, W.: Variable precision rough set model. J. Comput. Syst. Sci. **46**(1), 39–59 (1993). https://doi.org/10.1016/0022-0000(93)90048-2
29. Al-Najjar, N.I., Pai, M.M.: Coarse decision making and overfitting. J. Econ. Theory **150**, 467–486 (2014). https://doi.org/10.1016/j.jet.2013.12.003
30. Tversky, A., Kahneman, D.: Judgment under uncertainty: heuristics and biases. Science **185**(4157), 1124–1131 (1974). https://doi.org/10.1126/science.185.4157.1124
31. Simon, H.A.: A behavioral model of rational choice. Q. J. Econ. **69**(1), 99–118 (1955). https://doi.org/10.2307/1884852
32. Mohlin, E.: Optimal categorization. J. Econ. Theory **152**, 356–381 (2014). https://doi.org/10.1016/j.jet.2014.03.007
33. Mandler, M.: Coarse, efficient decision-making. J. Eur. Econ. Assoc. **18**(6), 3006–3044 (2020). https://doi.org/10.1093/jeea/jvaa002

34. Fagin, R., Halpern, J.Y., Moses, Y., Vardi, M.Y.: Common knowledge revisited. Ann. Pure Appl. Log. **96**(1), 89–105 (1999). https://doi.org/10.1016/S0168-0072(98)00033-5
35. Nash, J.: Non-cooperative games. Ann. Math. **54**(2), 286–295 (1951). https://doi.org/10.2307/1969529
36. Nash, J.F.: Equilibrium points in N-person games. Proc. Natl. Acad. Sci. USA **36**(1), 48–49 (1950). https://doi.org/10.1073/pnas.36.1.48
37. Howard, N.: Paradoxes of rationality: theory of metagames and political behavior. Cambridge, Mass.: MIT Press, 1971, 248 pp. Am. Behav. Sci. **15**(6), 948 (1972)
38. Fraser, N.M., Hipel, K.W.: Solving complex conflicts. IEEE Trans. Syst. Man. Cybern. **9**(12), 805–816 (1979). https://doi.org/10.1109/TSMC.1979.4310131
39. Fraser, N.M., Hipel, K.W.: Conflict Analysis: Models and Resolutions (North-Holland Series in System Science and Engineering). Elsevier Science Ltd., Amsterdam (1984)
40. Rapoport, M., Guyer, A.: A Taxonomy of 2×2 games. Gen. Syst. **11**, 203–214 (1966)
41. Hipel, K.W., Fang, L., Kilgour, D.M., Kilgour, M.: Environmental conflict resolution using the graph model. IEEE Syst. Man Cybern. (1993). https://doi.org/10.1109/icsmc.1993.384737
42. Kilgour, D.M., Hipel, K.W., Peng, X., Fang, L.: Coalition analysis in group decision support. Group Decis. Negot. (2001). https://doi.org/10.1023/A:1008713120075

Comparing Algorithms for Fair Allocation of Indivisible Items with Limited Information

Fahimeh Ziaei and D. Marc Kilgour(✉)

Department of Mathematics, Wilfrid Laurier University, Waterloo, Canada
ziae0890@mylaurier.ca, mkilgour@wlu.ca

Abstract. In studies of collective decision-making, the problem of allocating indivisible items fairly and efficiently is now recognized as the most difficult. Here, various algorithms for finding allocations are assessed on their ability to achieve the desirable properties of envy-freeness, Pareto-optimality, maximin, maximum Borda sum, and Borda maximin. Two players with additive preferences allocate an even number of indivisible items when their only information is the other's strict preference ordering. Algorithms under study include both naive and sophisticated versions of sequential selection, bottom-up sequential selection (or sequential rejection), balanced alternation, bottom-up balanced alternation, and fallback bargaining. The results suggest that fallback bargaining, the only simultaneous algorithm, satisfies most fairness and efficiency criteria but has some distinctive drawbacks.

Keywords: 2-person fair division · indivisible items · Pareto-optimal · envy-free

1 Introduction

The study of fair allocation addresses a wide range of problems from complete information about preferences and utilities to lack of information or uncertainty about the values of others. Fairly allocating divisible items may be possible when full information about utilities is available, but the challenge increases dramatically when the resources to be allocated consist of indivisible items. In our earlier paper [1] we considered how four indivisible items could be allocated to two players (two items each) so as to satisfy criteria of envy-freeness, Pareto-optimality, maximin, maximum Borda sum, and Borda maximin. In this paper broaden our assessment of fair division of indivisible items to two players and any even number of items.

Research on fair-division problems and their applications includes Thomson et al.'s general theory of fair allocations [2]. The general framework for modeling a set of players and indivisible items, where each player's preferences over the items are given as utilities, is debated by Sen [3] and Kilgour and Vetschera

Y. Maemura et al. (Eds.): GDN 2023, LNBIP 478, pp. 130–141, 2023.
https://doi.org/10.1007/978-3-031-33780-2_9

[4]. Some studies focus on situations in which participants have complete knowledge of each other's preferences over items and compare algorithms, aiming to attain three properties of fairness and efficiency: envy-freeness, Pareto optimality, and max-min [5–7]. Voting procedures may help, but sometimes no envy-free allocation exists [8].

This paper aims to compare algorithms for the balanced allocation of indivisible items—that is, each player receives the same number of items. The only information available to the algorithm is each player's preference ordering of the items. (Each player is assumed to have additive utilities, but only the ordering of the utilities is known.) Insofar as possible, we find good allocations using both top-down (select for yourself) and bottom-up (reject and impose on the other) approaches. The basic criteria of Pareto-optimality, envy-freeness, and maximin measure the efficiency and fairness of allocations [9]. Borda properties, based on Borda counts, are also used to assess allocations [10]. Many practical problems of allocation in the real world are exemplified by two-player problems, which must be solved prior to addressing more ambitious applications, like the allocation of portfolios of assets, the drafting of players to sports teams, and the division of assets in a divorce, to name but a few.

2 Properties of Allocations

In this paper, two players must share a set of finite even number of indivisible items that each strictly ranks from best to worst. We first set out the main assumptions of all models used in this paper:

1. **Strict preferences.** Each player's preference over the individual items forms a strict ordering.
2. **Self-interest.** Each player aims to obtain the best items it can, not to hurt the opponent.
3. **Independence.** Each player acts independently. There are no coalitions or hidden agreements.
4. **Partial Information.** Both players know each other's preference orderings, though not each other's utilities.
5. **Synergy-free.** There are no synergies, positive or negative, among the items than any player may receive.

Consider a set S of items to be allocated to players A and B. Suppose $X_A \subseteq S$ and $X_B \subseteq S$ satisfy $X_A \cap X_B = \emptyset$ and $X_A \cup X_B = S$. Then $X = (X_A, X_B)$ is the allocation in which X_A is assigned to A and subset X_B is assigned to B. We consider only balanced allocations, in which $|X_A| = |X_B|$.

We assume that players have preferences on S. For player $m \in M = \{A, B\}$, $x \prec_m y$ means that item $x \in S$ is less preferred by m than item $y \in S$. In this case, we also write $y \succ_m x$. We assume that player m's preferences are complete, asymmetric, and irreflexive, and therefore form a linear (strict) ordering. Player m's rank for item $x \in S$ is $r_m(x)$, the number of items in S that m prefers to x. Each player's preference ordering on S is assumed to be public information.

We assume that players have preferences on subsets of S, which are needed to assess allocations. To some extent, those preferences are implied by the player's preference ordering of individual items. We say that $X \subseteq S$ is ordinally less than $Y \subseteq S$ for m, denoted $X \prec_m Y$, if there exists an injective mapping $f : X \to Y$ so that $\forall x \in X$, $x \prec_m f(x)$. If $X \prec_m Y$, then $Y \succ_m X$, and we say that Y is ordinally more than X. But we emphasize that each player has some preferences between subsets that cannot be captured using the ordinally less criterion.

In all of our examples, we name the items in order of A's preference, so that A's preference ordering of S is always

$$A: \ 1 \succ_A 2 \succ_A \ldots \succ_A n$$

Of course, there are $n!$ possible orders of preference for B, so that there are $n!$ distinct problems with n items.

If $T \subseteq S$, $T \neq \emptyset$, let $\min_m\{T\}$ be player m's least preferred item in T. Thus, $\min_m\{T\} = x$ if and only if $x \in T$ satisfies $y \succ_m x$ for all $y \in T - \{x\}$.

We now introduce three properties commonly used in the literature of fair division of indivisible items:

- **Pareto Optimality (PO)**: An allocation X is Pareto-optimal (PO) iff there is no other allocation Y such that, for both players, the assignment under X is less preferred than the assignment under Y.
- **Envy-Freeness (EF)**: An allocation X is envy-free iff each player prefers its assignment to the subset assigned to the opponent [9].
- **Max-Min property (MM)**: An allocation is max-min iff there is no other allocation in which the maximum rank of any item in any player's assignment is less.

Unfortunately, EF and PO cannot be used to define "optimality" because, no envy-free allocation exists in many problems, and in others there are envy-free allocations that are not PO. Thus, no procedure can always yield a Pareto-optimal envy-free allocation. This leads us to add "possibly envy-free" and "possibly Pareto-optimal" to our criteria for good allocations; these criteria take into account players' preferences on subsets beyond those captured in the ordinally less relation.

The Borda score of player $m \in M$ for an assignment $X_m \subseteq S$ is

$$R_m(X) = \sum_{x \in X_m} r_m(x)$$

The two Borda criteria [10] for algorithms are

- **BordaSum (BS)**: A balanced allocation X is BordaSum iff there is no other balanced allocation in which the sum of the two players' Borda scores is less than $R_A(X) + R_B(X)$.
- **Borda Max-min (BM)**: An allocation X is Borda Max-min iff there is no other balanced allocation such that the maximum of two players' Borda scores is less than $\max\{R_A(X), R_B(X)\}$.

3 Algorithms

We study algorithms that the players can implement themselves, in the sense that no referee is required. Some, called "sophisticated," are explicitly game-theoretic, and assume far-sighted rationality; each of these is accompanied by a naïve version, in which players are short-sighted.

3.1 Sequential Algorithms

Sequential algorithms are implemented item-by-item; each step involves only player choosing or rejecting a specific item.

3.1.1 Sophisticated Sequential Selection

Sequential Selection is implemented in many professional sports when teams select (or "draft") new players in sequence; teams with poorer records in the previous season usually get to pick earlier [11]. Our model focuses on two teams, which we call players, A and B. We assume that player A chooses first, then B, who may choose any item other than the one that A selected. The process repeats until the available items are exhausted. Note that Player A has priority in every round.

Sophisticated Sequential Selection is an algorithm that assumes that each player always chooses optimally and expects that all subsequent choices will be optimal. Study of the process shows that Sophisticated Sequential Selection always produces an outcome that is balanced and Pareto-optimal, but not necessarily envy-free, Max-Min, BordaSum, or Borda Max-min.

The Kohler allocation, determined using Kohler numbers, is a balanced allocation that could result from Sophisticate Sequential Selection. Each item in $p \in S$ has a well-defined Kohler number, K_p, where $1 \leq K_p \leq n$; each Kohler number is unique. After the Kohler numbers have been assigned, it is easy to find the Kohler allocation:

$$S_A = \{K_1, K_3, \ldots, K_{n-1}\}; \qquad S_B = \{K_2, K_4, \ldots, K_n\}$$

The following procedure is used to determine Kohler numbers:

Identify K_n
Define $S_n = S$ and

$$K_n = \min_A\{S_n\}$$

Set $S_{n-1} = S_n - \{K_n\}$.

Identify K_p *for* $p = n-1, n-2, \ldots, 2$
Assume $K_n, K_{n-1}, \ldots, K_{p+1}$ have been assigned, and that $|S_n| \geq 2$. Then let $X = A$ if p is even and $X = B$ if p is odd. Define

$$K_p = \min_X\{S_p\}$$

Set $S_{p-1} = S_p - \{K_p\}$. If $|S_{p-1}| = 1$, then assign K_1 to be the unique element of S_{p-1}, completing the determination of Kohler numbers. Otherwise, reduce p by 1 and repeat this step.

Example 1: $n = 4$

$$A : 1 \succ_A 2 \succ_A 3 \succ_A 4$$

$$B : 2 \succ_B 4 \succ_B 3 \succ_B 1$$

The Kohler numbers are
$K_4 = 4$, which is A's least preferred item in $S = S_4 = \{1, 2, 3, 4\}$
$K_3 = 1$, which is B's least preferred item in $S_3 = \{1, 2, 3\}$
$K_2 = 3$, which is A's least preferred item in $S_2 = \{2, 3\}$
$K_1 = 2$, since $|S_1| = |\{2\}| = 1$
Then the Kohler allocation is

$$S_A = \{K_1, K_3\} = \{2, 1\}, \quad S_B = \{K_2, K_4\} = \{3, 4\}$$

The extensive-form game shown in Fig. 1 represents the sequential selection process. If all choices are rational under the expectation that all future choices will be rational in the same sense, then they form a subgame-perfect equilibrium (SPE) of this game.

In Fig. 1, the arcs coloured red or blue are determined by backward induction. All of these arcs are consistent with SPE. Because of our assumption that the players' preferences are strict, each player always has an optimal choice at each of its nodes. For example, in the node at the top left, by choosing 2, B obtains

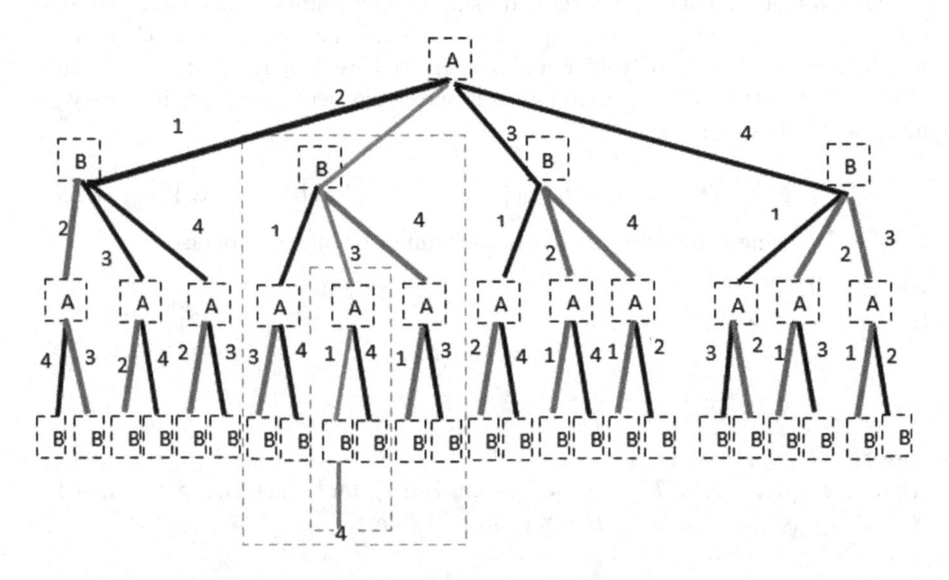

Fig. 1. Sophisticated Selection in Example 1: Game Tree

outcome $\{2,4\}$, which it prefers to outcome $\{3,4\}$, which would follow from the choice of 3 or 4. B prefers 24 to 34 because B prefers 2 to 3. (At some nodes, a player has more than one optimal choice, as more than one arc leads to the most preferred outcome achievable from that node.)

Note that in Fig. 1 there are two paths consisting only of red or blue arcs from the initial node to a terminal node. Both of these paths, (2, 3, 1, 4) and (2, 4, 1, 3), represent SPEs, and there are no other SPEs. Both paths lead to the outcome (12, 34), which is therefore the outcome of sophisticated (or "rational") play in this game. To achieve it, A must begin by choosing 2; then B is indifferent between 3 and 4, either of which leads to same allocation, $S_A = \{1,2\}; S_B = \{3,4\}$ (Table 1).

Table 1. Allocation Sequence in Example 1

	Round 1		Round 2	
	Item	Kohler number	Item	Kohler number
Player A	2	K_1	1	K_3
Player B	3	K_2	4	K_4

It is a remarkable fact that the Kohler algorithm actually produces an SPE, in this case the one indicated in blue in Fig. 1. [12] As Table 3 below shows, the Kohler algorithm identifies not only the subsets resulting from optimal selection, (12, 34), but also the order of selection. In the first round, A selects $K_1 = 2$ and then B selects $K_2 = 3$. In the second round, A selects $K_3 = 1$ and then B is left with $K_4 = 4$.

Example 2: $n = 8$

$$A : 1 \succ_A 2 \succ_A 3 \succ_A 4 \succ_A 5 \succ_A 6 \succ_A 7 \succ_A 8$$

$$B : 7 \succ_B 2 \succ_B 8 \succ_B 5 \succ_B 4 \succ_B 1 \succ_B 3 \succ_B 6$$

The Kohler numbers for Example 2 are
$K_8 = 8$, which is A's least preferred item in $S = S_8 = \{1,2,3,4,5,6,7,8\}$
$K_7 = 6$, which is B's least preferred item in $S_7 = \{1,2,3,4,5,6,7\}$
$K_6 = 7$, which is A's least preferred item in $S_6 = \{1,2,3,4,5,7\}$
$K_5 = 3$, which is B's least preferred item in $S_5 = \{1,2,3,4,5\}$
$K_4 = 5$, which is A's least preferred item in $S_4 = \{1,2,4,5\}$
$K_3 = 1$, which is B's least preferred item in $S_3 = \{1,2,4\}$
$K_2 = 4$, which is A's least preferred item in $S_6 = \{2,4\}$
$K_1 = 2$, since $|S_1| = |\{2\}| = 2$
The Kohler allocation is

$$S_A = \{K_1, K_3, K_5, K_7\} = \{2,1,3,6\} \qquad S_B = \{K_2, K_4, K_6, K_8\} = \{4,5,7,8]$$

Table 2. Allocation Sequence in Example 2

	Round 1		Round 2		Round 3		Round 4	
	Item	Kohler	Item	Kohler	Item	Kohler	Item	Kohler
Player A	2	K_1	1	K_3	3	K_5	6	K_7
Player B	4	K_2	5	K_4	7	K_6	8	K_8

It can be proved that the Kohler method always identifies a subgame perfect equilibrium of the extensive-form representation of a selection problem with an even finite number of items [12] (Table 2).

3.1.2 Naive Sequential Selection

Naïve outcomes occur when the players act in a short-sighted manner. The first three of our assumptions are important here—strict preferences, self-interest, and independence—but the others are not relevant. We simply assume that each player always selects the most preferred available item.

We illustrate the process with Example 2, repeated here for convenience:

Example 2: $n = 8$

$$A : 1 \succ_A 2 \succ_A 3 \succ_A 4 \succ_A 5 \succ_A 6 \succ_A 7 \succ_A 8$$

$$B : 7 \succ_B 2 \succ_B 8 \succ_B 5 \succ_B 4 \succ_B 1 \succ_B 3 \succ_B 6$$

The players select sequentially, always picking their most preferred available items, with player A starting. The result is

$$S_A = \{1, 2, 3, 4\} \quad S_B = \{7, 8, 5, 6\}$$

Note that Naïve Sequential selection produces a Pareto-optimal outcome, but the outcome is not necessarily envy-free or Max-Min. Also, BordaSum and Borda Max-min can also fail under this algorithm.

The proof that an allocation achieved by Naive Sequential Selection is Pareto-Optimal is similar to the proof for Sequential Sophisticated Selection [13]. The reason is that it follows the rule of identifying the Kohler numbers from up to down, while the Sophisticated selection identifies the Kohler numbers from bottom to top.

3.1.3 Sophisticated Sequential Rejection

So far, we analyzed methods in which players choose items they want. Now we turn to methods in which players reject items they do not want, forcing the opponent to accept them. These methods are also known as Bottom-Up Sequential methods.

We illustrate Sophisticated Sequential Rejection with Example 2, repeated here for convenience:

Example 2: $n = 8$

$$A : 1 \succ_A 2 \succ_A 3 \succ_A 4 \succ_A 5 \succ_A 6 \succ_A 7 \succ_A 8$$

$$B : 7 \succ_B 2 \succ_B 8 \succ_B 5 \succ_B 4 \succ_B 1 \succ_B 3 \succ_B 6$$

Player A starts the game. Item 8 is the item A wants the least, but this item is among the more preferred for player B, so A does not reject it. The same logic applies for item 7. So, player A rejects item 6, forcing it on player B. Player B, whose goal is to benefit itself by assumption (2), knows that player A will never reject items 1 or 3, among his favorites, so player B rejects item 4. In the second round, player A rejects item 5 and player B rejects item 3. In the third round, the wise strategy for player A is to reject item 8, and player B will reject item 2, since he knows that item 1 is player A's most favorite item. In the round 4, player A will reject item 7 and player B has no defense against this strategy and can do no better than reject item 1. The resulting choices are then

$$S_A = \{1, 2, 3, 4\} \quad S_B = \{7, 8, 5, 6\}$$

which is different from the result of Sophisticated Sequential Selection in the same example ($S_A = \{2, 1, 3, 6\}$, $S_B = \{4, 5, 7, 8\}$). Player A receives better results with Sophisticated Sequential Rejection than with Sophisticated Sequential Selection, but player B has the opposite preference.

These methods are also known as Bottom-Up Sequential methods. The analysis of the sophisticated actions of players can be shown by a game tree (similar to Fig. 1, for Sophisticated Sequential Selection). The results show that this algorithm always produces a Pareto-optimal outcome—no other allocation would make one participant better off without making the other worse off. The proof is the same as the Sequential Naive Selection method, due to the fact that both methods produces the same outcomes. This algorithm does not always produce outcomes that satisfy the properties of envy-freeness, Maximin, BordaSum, or Borda Max-min.

3.1.4 Sequential Naïve Rejection

This method is based on the sincere sequential rejections of players over the items. The players act short-sightedly, and sincerely reject their least favored items. This method is also Pareto Optimal, since it produces the same results with the Sequential Naive selection(this uniformity of results has the roots in the Kohler algorithm of crossing out the players sequentially from the bottom to up in the Sophisticated Method), but not envy-free, Max-Min, BordaSum, or Borda Max-min.

3.1.5 Sophisticated Balanced Selection

Sequential selection is based on taking turns: player A picks an item; then player B picks one; then Player A chooses again; and so on. Of course, going first can be a huge advantage. Giving extra choices to compensate for going second can reduce, if not eliminate, this advantage. A specific way of balancing choices yields a procedure called Balanced Alternation [14]. If, for instance, the order (A, B) in the first two rounds gives an advantage to player A, then reversing the order in rounds three and four (B, A) will tend to compensate. If there are more than four items, the same principle applies recursively for the next 4 (or the next $2n$) rounds, producing sequences like (A, B, B, A, A, B, B, A), etc.

We illustrate Sophisticated Balanced Rejection with Example 2, repeated here for convenience:

Example 2: $n = 8$

$$A : 1 \succ_A 2 \succ_A 3 \succ_A 4 \succ_A 5 \succ_A 6 \succ_A 7 \succ_A 8$$

$$B : 7 \succ_B 2 \succ_B 8 \succ_B 5 \succ_B 4 \succ_B 1 \succ_B 3 \succ_B 6$$

In this example, player A starts the game, knowing that the first item is not among the preferred items of player B, so the wise choice for him it to choosing 2, then player B will choose items 5 and 4, since items 7 and 8 are not among favorite items of player A. Player A will choose item 1. In the next round, player A will choose 3, and player B has no option better than choosing 7 and 8, which leave the item 6 to player A. Therefore, the allocation would be $S_A = \{2, 1, 3, 6\}$ and $S_B = \{5, 4, 7, 8\}$. This method always produces possibly Pareto Optimal and possibly envy-free allocations but the allocation may be Max-Min, BordaSum, or Borda Max-min. Kohler number cannot be defined for this method since we cannot identify and assign the K's to items according to Minimum preference (or Maximum preference) set. But because the players play according to their preference orderings implies that the result is always possibly Pareto Optimal.

3.1.6 Naive Balanced Selection

This main logic of this method is the same as Sophisticated Balanced Selection. Players always pick the most preferred remaining item. Note that our 4th assumption is not applied here since complete information about the opponent's preferences does not affect a player's decision-making.

To illustrate Naive Balanced Rejection, we apply it to Example 2, producing the allocation $S_A = \{1, 3, 4, 6\}$ and $S_B = \{7, 2, 8, 5\}$. This method always produces a possibly Pareto optimal and possibly envy-free allocation, which may not satisfy the Maximin, BordaSum, and Borda Max-min criteria.

3.1.7 Sophisticated Balanced Rejection

This method works based on the rejection of unwanted items, but according to the Balanced Alternation principle. Thus, the first player can reject one item, and then the second player can reject two items, then the first can reject two, then the second, and the game continues in this manner. This method is possibly Pareto-optimal and possibly envy-free, but may not be maximin, BordaSum, and Borda Max-min.

3.1.8 Naive Balanced Rejection

This method works based on rejecting the lowest-ranked items of the player's preference list in a balanced manner according to each player's short-term vision. This method is also possibly Pareto optimal and possibly envy-free, but not maximin, BordaSum, and Borda Max-min.

3.2 Simultaneous Algorithm

Until now, the algorithms introduced based on item-by-item selection. Now we introduce an algorithm that finds a suitable settlement by considering simultaneously all possible allocations. The player must reveal at the outset their complete preference ranking of all possible balanced allocations, rather than making through item-by-item choices. This procedure requires a neutral referee (which could be a computer program).

3.2.1 Fallback Bargaining

Fallback bargaining method is a step-by-step procedure to determine a 'compromise allocation,' carried out as follows [15]:

(1) Consider each player's most preferred allocation. If it is the same for both players, then it is implemented (depth 1 agreement) and the process stops.
(2) If there is no common agreement after j steps, both players' next-most-preferred allocations (ranked $j+1$) are considered. If there is an allocation within each player's top $j+1$ choices, it is implemented, and constitutes a depth $j+1$ agreement. Otherwise, increase j by 1 and repeat Step (2).

Note that there is a finite number, j, such that at least one allocation is common to both players' top j allocations, while no allocation is common to their top $j-1$ allocations. Any such allocation is a depth j agreement. One important drawback of the Fallback Bargaining procedure is that two allocations may tie, both becoming mutually acceptable at the same level. Another drawback is that the final agreement may require a very long process if n is large. If their preferences are exactly opposite, agreement will require $\frac{1}{2}\binom{n}{\frac{n}{2}}$ steps.

The Fallback Bargaining method is possibly Pareto-optimal, possibly envy-free, Max-Min, and Borda Max-min. It fails the BordaSum criterion.

4 Results

Our main research question was how to find a good balanced allocation of any even number of indivisible items based only two players' preference orderings, where a good allocation is defined using the criteria of PO, EF, and MM as well as BS and BM. We considered only algorithms that players could operate without aid in the sense of game theory, and assessed whether they achieved desirable properties for certain, no matter what the players' preference orderings.

Our findings are summarized in Table 3. The only algorithm that can be relied on to achieve most of the criteria of efficiency and fairness is Fallback Bargaining. We obtained the same result for our previous analysis of 4-item allocation problems; we have shown that it extends to any even number of items. To explain it, note that sequential selection allocations always give an advantage to player A, who starts the game and can guarantee receipt of the most preferred item (or rejection of the least preferrred). Player A may not take advantage of this option on the first move, but only because he is confident that his first priority will eventually be achieved. This advantage ensures that sequential selection allocations are Pareto optimal, but not envy-free or Max-min since, in many cases, player B would also prefer what A can guarantee for himself.

Balanced Alternation returns at least some of A's advantage to B. The consequence is that allocations are not always Pareto-optimal but are always possibly Pareto-optimal. Allocations determined by this procedure are always possibly envy-free, but not necessarily Max-min.

In the Fallback Bargaining algorithm, the players act simultaneously, which is the key to the algorithm's ability to find an allocation that satisfies most of the criteria of efficiency and fairness, including Borda Maximinality. Unfortunately, it has larger information requirements, fails uniqueness, and may take many steps to reach a conclusion.

Table 3. Properties of the Algorithms

		Method	unique result	Envy-Free	Possibly Envy-Free	Pareto Optimal	Possibly Pareto-optimal	Max-Min	Borda Sum	Max-min Borda
Item-By-Item	Sequential	Sophisticated Selection	Yes	No	No	Yes	No	No	No	No
		Naive Selection	Yes	No	No	Yes	No	No	No	No
		Sophisticated Rejection	Yes	No	No	Yes	No	No	No	No
		Naive Rejection	Yes	No	No	Yes	No	No	No	No
	Balanced	Sophisticated Selection	Yes	No	Yes	No	Yes	No	No	No
		Naive Selection	Yes	No	Yes	No	Yes	No	No	No
		Sophisticated Rejection	Yes	No	Yes	No	Yes	No	No	No
		Naive Rejection	Yes	No	Yes	No	Yes	No	No	No
Simultaneous		Fallback Bargaining	No	No	Yes	No	Yes	Yes	No	Yes

References

1. Ziaei, F., Kilgour, D.M.: Comparing algorithms for fair division of indivisible items. In: 22th International Proceedings of Group Decision and Negotiation, GDN 2022, pp. 51–62 (2022)
2. Thomson, W.: Introduction to the theory of fair allocation. In: Brandt, F., Conitzer, V., Endriss, U., Lang, J., Procaccia, A.D. (eds.) Handbook of Computational Social Choice (2016)
3. Sen, A.: The possibility of social choice. Am. Econ. Rev. **89**, 349–378 (1999)
4. Kilgour, D.M., Vetschera, R.: Two-player fair division of indivisible items: comparison of algorithms. Eur. J. Oper. Res. **271**(2), 620–631 (2018)
5. Brams, S.J., Kilgour, D.M., Klamler, C.: Maximin envy-free division of indivisible items. Group Decis. Negot. **26**(1), 115–131 (2017)
6. Pruhs, K., Woeginger, G.J.: Divorcing made easy. In: Kranakis, E., Krizanc, D., Luccio, F. (eds.) FUN 2012. LNCS, vol. 7288, pp. 305–314. Springer, Heidelberg (2012). https://doi.org/10.1007/978-3-642-30347-0_30
7. Bouveret, S., Lemaître, M.: Characterizing conflicts in fair division of indivisible goods using a scale of criteria. Auton. Agent. Multi-Agent Syst. **30**(2), 259–290 (2016)
8. Lang, J.: Collective decision making under incomplete knowledge: possible and necessary solutions. In: 29th International Joint Conference on Artificial Intelligence and Seventeenth Pacific Rim International Conference on Artificial Intelligence, Yokohama, Japan, pp. 4885–4891 (2020)
9. Klamler, C.: The notion of fair division in negotiations. In: Handbook of Group Decision and Negotiation (2021)
10. Darmann, A., Klamler, C.: Proportional Borda allocations. Soc. Choice Welfare **47**(3), 543–558 (2016). https://doi.org/10.1007/s00355-016-0982-z
11. Brams, S.J., Straffin, P.D., Jr.: Prisoners' dilemma and professional sports drafts. Am. Math. Mon. **86**(2), 80–88 (1979)
12. Kohler, D.A., Chandrasekaran, R.: A class of sequential games. Oper. Res. **19**(2), 270–277 (1971)
13. Brams, S.J., King, D.L.: Efficient fair division: help the worst off or avoid envy? Ration. Soc. **17**(4), 387–421 (2005)
14. Brams, S.J., Taylor, A.D.: Fair Division: From Cake-Cutting to Dispute Resolution. Cambridge University Press, Cambridge (1996)
15. Brams, S.J., Kilgour, D.M.: Fallback bargaining. Group Decis. Negot. **10**(4), 287–316 (2001)

Hierarchical Modeling of Aggregate Mining Conflict in Ontario, Canada

Nayyer Mirnasl[1](✉), Keith W. Hipel[1,2,3], Simone Philpot[1], and Aidin Akbari[4]

[1] Department of Systems Design Engineering, University of Waterloo, Waterloo, ON, Canada
snmirnas@uwaterloo.ca

[2] Centre for International Governance Innovation, Waterloo, ON, Canada

[3] Balsillie School of International Affairs, Waterloo, ON, Canada

[4] Waterloo, ON, Canada

Abstract. To develop strategic insights regarding the aggregate mining disputes in the province of Ontario, Canada, and to better understand their structured power hierarchies, we developed a two-level hierarchical graph model based on three case-specific classical conflict models. The stability analysis of the two-level hierarchical graph model shows that despite producing potential resolutions at the local level (sub-models), the model does not result in an equilibrium. This finding suggests that in the absence of preferences by the common decision-makers (i.e., provincial, and local governments), finding potential resolutions to this class of disputes may be difficult.

Keywords: Aggregate mining · conflict resolution · Hierarchical graph model · Two-level Hierarchical Graph model · Ontario

1 Introduction

Aggregates refer to a family of particulate raw materials, including gravel, sand, clay, limestone, marble, granite, and similar coarse-to-fine grained raw resources extracted from pits and quarries [1]. Widely used in the production of composite materials such as asphalt and concrete, aggregates are the most mined raw resources in the world [2–5]. Whether small or large in scale, extraction of minerals often involves the use of heavy machinery, explosive materials, and open-pit mining techniques, all with adverse environmental impacts on water, land, and soil resources [3, 6–8].

Over the past several decades, the diminishing availability and growing demand for aggregate resources have brought about fundamental challenges concerning their sustainable use, particularly in or near ecologically sensitive areas [9, 10], which are sometimes located near human settlements. The result has been a growing trend in the number of conflicts over competing values and between different stakeholders [11]. In 2019, there was a total of 6105 aggregate licenses and permits, of which 3,614 licenses were for pits and quarries on private lands, and 2,491 were for pits and quarries on Crown lands [12].

A. Akbari—Independent Researcher.

Y. Maemura et al. (Eds.): GDN 2023, LNBIP 478, pp. 142–160, 2023.
https://doi.org/10.1007/978-3-031-33780-2_10

One of the areas with the most recurring instances of aggregate-related conflicts, particularly since the 1970s, is Ontario [13], Canada's most populous province and its largest economy [14, 15]. Consistent population growth and the need for further development, particularly across southern Ontario, have led to increased aggregate mining, often encroaching on areas with prime agricultural lands or other valued environmental resources [13, 16, 17]. Most land use planning scenarios associated with aggregate development proposals are regarded with disdain by the public and local stakeholders directly affected by these proposals [18, 19]. As a result, the number of conflicts involving municipalities, provincial governments, the aggregate industry, concerned environmental groups, and exasperated citizens has been on the rise [19].

Few studies are exploring aggregate issues and conflicts in this province [20–27], particularly in the southern parts of the province where approximately 80 percent of provincial mineral aggregate resources are produced [28] (Fig. 1).

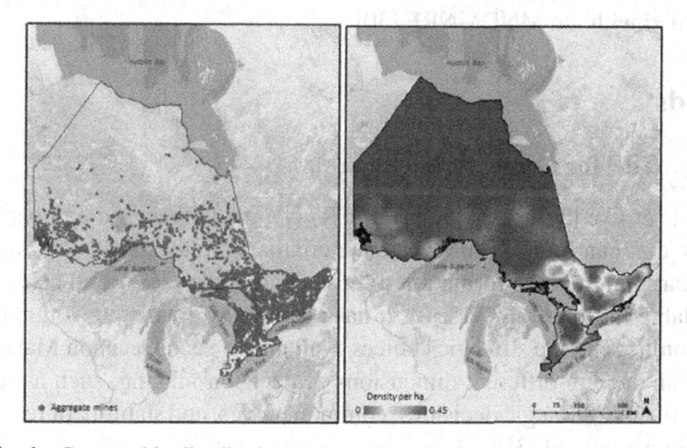

Fig. 1. Geographic distribution and density of aggregate mines in Ontario.

Building on three graph models of aggregate mining conflicts explored through the classical Graph Model for Conflict Resolution (GMCR), we develop a hierarchical graph model of aggregate mining conflicts in Ontario. The aim is to create a generic model which provides a broader understanding of this class of conflicts in the province.

2 Aggregate Mining Regulatory Framework in Ontario, Canada

Ontario's Ministry of Northern Development, Mines, Natural Resources and Forestry (MNDMNRF) governs all mining activities in the province. The ministry ensures that aggregate mining activities comply with the Aggregate Resources Act and other relevant legislation, such as the Planning Act, the Ontario Water Resources Act, the Environmental Protection Act, and the Endangered Species Act [27]. As required by the Environmental Bill of Rights, MNDMNRF also facilitates public participation in this and other significant environmental decisions in the province. Accordingly, the ministry must give residents at least sixty days to submit their comments on any aggregate mining license and permit applications [28].

Responsibilities related to detailed evaluation and approval of land use changes required for aggregate mines rest with regional and local municipalities in the province. Nevertheless, the provincial government has the power to override local governments' decisions by issuing a Minister's Zoning Order (MZO), an executive order issued by the Ministry of Municipal Affairs and Housing (MMAH) which prevails over municipal zoning bylaws [29].

In Ontario, the process involved in applying to excavate aggregates varies depending on a range of factors, including the type of land (private ownership or Crown land), the depth of extraction, the amount of aggregate to be extracted, and other details specific to a given application [1]. However, the general process requires (i) an official plan amendment application filed with the local municipality, county, or regional government when the proposed aggregate site has a different land use designation than what is listed on the official plan; (ii) a zoning bylaw amendment application filed with the municipality whenever there is a conflict with the existing zoning bylaws; and (iii) an aggregate license application filed with the MNDMNRF [30].

3 Methods

3.1 Graph Model for Conflict Resolution

Graph Model for Conflict Resolution (GMCR) is a game-theoretic approach based on graph theory that can be used to investigate conflicts strategically. This methodology utilizes ordinal preference information to investigate real-world conflicts based on the readily available information [31, 32]. It has a simple structure showing the state of a real-world conflict and the strategic choices available to each Decision Maker (DM). In addition, it can capture different dimensions of decision-making, such as reversibility (or irreversibility) of strategic decisions, common moves, and stability of resulting states for all DMs in a conflict [33].

The information needed to build a graph model includes DMs, options each DM can unilaterally control, a set of feasible states that could occur as the conflict unfolds, and the relative preferences of each DM over each state [31, 34–36]. The reachable list for a DM is a set of all states the DM can move to in one step from a pre-determined starting point in the conflict. These moves are known as Unilateral Moves (UMs). They become Unilateral Improvement (UI) each time a UM results in a more preferred state. That said, a graph model is a 4-tuple set containing DMs, states, arcs of moves, and preference relations. Mathematically, it can be written as $G = \{N, S, A, \succsim\}$ where $N = \{1, 2, \ldots, n\}$, S, $A = \{A_1, A_2, \ldots, A_n\}$, and $\succsim = \{\succsim_1, \ldots, \succsim_n\}$ respectively represent the set of DMs, the set of states, the profile of UMs, and the profile of preferences on S. For each $i \in N$, $A_i \subseteq S \times S$ is the set of all of i's UMs; $(s, s') \in A_i$ indicates that i can move (unilaterally and in one step) from state $s \in S$ to state $s' \in S$ [33, 36]. The notation $\succsim$ in this mathematical representation includes all possible relations between a pair of states, including more preferred ($\succ$), equally preferred ($\backsim$), less preferred ($\prec$), more or equally preferred ($\succsim$), and less or equally preferred ($\precsim$) [29].

Stability definitions—commonly referred to as solution concepts—are mathematical descriptions of different types of human behavior (DMs) in strategic conflicts. To identify which states are likely to be the resolutions of the conflict, states are assessed for their stabilities [36]. In stability analysis, moves and countermoves available to each DM are modeled alongside the preferences of each DM. Common solution concepts analyzed in GMCR include Nash stability (R) [37, 38], sequential stability (SEQ) [39], general metarationality (GMR) [40], and symmetric metarationality (SMR) [40], each reflecting different level of foresight, knowledge of information, and attitudes towards risk for DMs [41]. For each DM, a state can be stable under different solution concepts, but only states that are stable for all DMs are considered possible resolutions to the conflict. These potential resolutions are known as equilibria [36].

3.2 Hierarchical Graph Model

Sometimes real-world conflicts consist of smaller strategic interactions (games) between DMs. These linked conflicts are called hierarchical conflicts. Failure to perceive these hierarchical relations may lead to inaccurate inferences about strategic outcomes [42, 43]. This hierarchical structure has been widely discussed within the game theory paradigm [44] and graph theory [45–47].

The Hierarchical Graph Model for Conflict Resolution (HGMCR), an extension to classical GMCR, is an advanced approach to analyze interrelated conflicts with hierarchical structures [48]. In HGM, smaller graphs (known as local graphs) represent sub-conflicts, and DMs are classified into two main categories: a) Common DMs (CDMs) who are involved in all smaller games, and b) Local DMs (LDMs) who are only participants of one smaller game. DMs can initiate different types of moves and have interrelated preferences [49]. Preferences for CDMs are determined by assessing the preferences in all local graphs. For a CDM, a local graph can be more important ($\succ$), less important ($\prec$), or equally important ($\backsim$) to another one. States are the combination of all possible strategies in local games; strategic moves represent a collection of possible actions for DMs in sub-conflicts. Stabilities can be partially obtained from the stability calculations in local models [48–50].

3.3 Two-Level Hierarchical Graph Model

HGM presents a flexible structure; model structure can be modified and adapted to reflect real-world situations (e.g., CDMs and the type of interaction between them). For example, two-level Hierarchical Graph Models (2LHGMs) were developed to reflect situations in which DMs interact at two levels [48–51]. The structure of a 2LHGM is depicted in Fig. 2. The nodes in this graphical illustration represent CDMs and LDMs.

2LHGM comprises two Basic Hierarchical Graph Models (BHGMs), consisting of two sub-conflict models between two players: a CDM and an LDM. At the top level, a sub-conflict takes place between the two CDMs, i_1^1 and i_2^1. The names of the DMs in the 2LHGM are denoted by i_x^k, where k is the level, and x is each DM's position on its level [48, 50].

Mathematically, a 2LHGM, G^H, can be defined as:

If $G_1^2 =< \{i_1^1, i_{11}^2, i_{12}^2\}, S_1^2, A_1^2, \succsim_1^2\} >$ and $G_2^2 =< \{i_2^1, i_{21}^2, i_{22}^2\}, S_2^2, A_2^2, \succsim_2^2\} >$ are two BHGMs and $G_0^1 =< \{i_1^1, i_2^1\}, S_0^1, A_0^1, \succsim_0^1\} >$ is the local graph at the top level

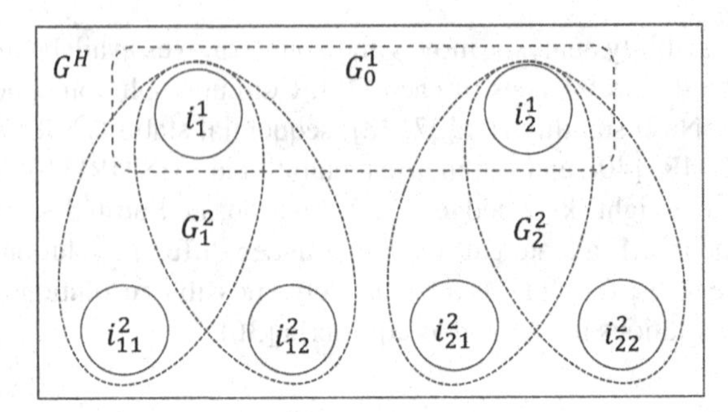

Fig. 2. Two-Level Hierarchical Graph Model (2LHGM) conceptual diagram [50]

containing two CDMS i_1^1 and i_2^1; then, a 2LHGM is defined as $G^H = G_0^1 \cup G_1^2 \cup G_2^2 =< N^H, S^H, A^H, \succsim^H >$ where $N^H = \left\{i_1^1, i_2^1, i_{11}^2, i_{12}^2 i_{21}^2, i_{22}^2\right\}$ and $S^H \subseteq S_0^1 \times S_1^2 \times S_2^2$.

$s^H \in S^H$ in G^H is Nash stable fori_1^1, a CDM, if s_0^1 and s_1^2 are Nash stable in G_o^1 and G_1^2 for i_1^1, respectively. And $s^H \in S^H$ is Nash stable for i_{11}^2, an LDM, if and only if s_1^2 is Nash stable in G_1^2 for i_{11}^2. Also, $s^H \in S^H$ is SEQ for i_1^1, a CDM, in G^H if and only if s_0^1 is SEQ stable for i_1^1 in G_o^1 or s_0^1 is SEQ stable for i_1^1 in G_o^1 and s_1^2 is SEQ for i_1^1 in G_1^2. Also, $s^H \in S^H$ is considered SEQ for i_{11}^2, an LDM, if and only if s_1^2 is SEQ stable in G_1^2 for i_{11}^2 [48, 50].

The Nash and SEQ stabilities are the only solution concepts considered in this paper, as they portray typical behavioral patterns displayed by DMs in a real-world aggregate mining conflict. Put differently, no DM in an aggregate mining conflict would choose strategic disimprovement at one point in the conflict in response to other DMs' UIs to gain strategic advantage over them at a later point in time.

3.4 Description of Three Case Studies of Aggregate Mining Conflict in Ontario and Data Sources

Herein, three GMCR-based aggregate mining conflict analyses are used to develop a 2LHGM for aggregate mining conflicts in Ontario. These published studies were focused on conflicts arising from different aggregate mining applications in the province. The following provides a glance into each of these conflict studies.

The Hallman Pit conflict study focused on the conflict arising from the Hallman Pit application, a Class "A" license for aggregate extraction above the water table [52]. The application proposed zoning of 161 acres of land, change of licensing to pit quarry in 141 acres, and allowance for annual extraction volume of 750,000 tons of aggregate in 129 acres [53] with most designated areas identified as Prime Agricultural lands with Protected Countryside and Mineral Aggregate Area status [26].

The Campbellville Pit Conflict study analyzed a Category "A" pit/quarry application which would allow the extraction of 990,000 tonnes of aggregates below the water table for twenty years [24].

The Teedon Pit conflict study investigated the dispute arising over the Teedon Pit expansion application, which would directly affect the groundwater quality and recharge zone for Tiny Township's Alliston Aquifer [25].

The data used for these conflicts were obtained from publicly available documents (e.g., materials posted online at the Environmental Registry of Ontario) about each application, as well as other relevant data specific to each study collected from online newspapers, citizens, environmental and industry websites, and relevant legislation such as the Aggregate Resources Act and the Ontario Planning Act [24–26]. Table 1 summarizes the set of DMs and their respective options for each conflict case described above.

3.5 Application of 2LHGM on Aggregate Mining Conflict in Ontario

The need to construct a 2LHGM for aggregate mining conflicts in Ontario arises from a growing need to understand different power dynamics in this class of provincial disputes. More specifically, as informative as a classical GMCR would be for analyzing these conflicts, a one-level conflict model might partially fail to portray the existing structured hierarchies of power and some micro-level strategic interactions in this class of conflicts. That is, an option selected by a more powerful DM in a classical GMCR could render other DM's options ineffective and change the fate of the conflict instantaneously. A good example is the use of MZOs by the provincial government in these types of conflicts. Although this is not far from reality in many conflict situations, micro-level interactions between the otherwise inferior DMs (and sometimes more powerful ones) could have influenced the potential resolutions had they been considered in the analysis. By synthesizing the data and insights from the results of the three considered one-level GMCR studies, we collected all the required information to build a 2LHGM for aggregate mining conflicts in Ontario.

Figure 3 shows a conceptual diagram of a 2LHGM for aggregate mining conflict in Ontario and the interactions between DMs. It also illustrates the interactions between different DMs. There are two CDMs in the model. In the first level (G_0^1), Provincial Government and Local Government are involved in the game as the two DMs. In the second level, there are two BHGMs, namely G_1^2, G_2^2. The first BHGM, G_1^2, includes the Applicant and Ontario Land Tribunal as the two LDMs and Provincial Government as the only CDM. The second BHGM, G_2^2, also has two LDMs, Citizens and Applicant. In this game, Local Government is the CDM. The options of CDMs and LDMs in each game and their preferences are described in the following sections.

3.6 Decision-Makers and Their Options in 2LHGM

The DMs and their options for 2LHGM for aggregate mining conflicts in Ontario are listed in Table 2. Accordingly, the Applicant controls whether to pursue the license application. They may first need to apply for zoning bylaw changes. Based on the Aggregate Resources Act, local government has the power to change zoning bylaws. If the local government rejects the application, the applicant can then appeal to the Ontario Land Tribunal (hereafter Tribunal), the judicial body that adjudicates or mediates matters related to land use planning, environmental and heritage protection, land valuation, mining, and

Table 1. List of DMs and their options for each case in previous studies

Decision Makers	Options	Explanation*
Hallman Pit Conflict		
Jackson Harvest Farms (JHF)	1. Zoning Application	Y: Applies for zoning bylaw changes
		N: Does not apply to zoning bylaw changes
	2. LPAT Appeal	Y: Appeals decision
		N: Does not appeal decision
Local Government (LG)	3. Approve	Y: Approves land use changes
		N: Approval was not issued
	4. Reject	Y: Rejects the land use changes
		N: Approval was not issued
Local Planning Appeal Tribunal (LPAT)	5. Approve	Y: Directs the province to issue the aggregate license
		N: Rejects the request
	6. Conditions	Y: Directs the province to issue the aggregate license with additional conditions
		N: Does not add conditions to the application
	7. Reject	Y: Refuses to issue license
		N: Does not direct province to reject the license
Province	8. Approve	Y: Approves the license
		N: Does not approve the license
	9. Conditions	Y: Approves the license with additional conditions
		N: Does not approve the license
Citizens for Safe Groundwater (CSGW)	10. LPAT Appeal	Y: Appeals decision
		N: Does not appeal decision
Campbellville Quarry Conflict		
Ministry Of Environment, Conservation and Parks (MECP)	1. Approve	Y: Approves the license N: Rejects the license application

(*continued*)

Table 1. (*continued*)

Decision Makers	Options	Explanation*
	2. Conditional Approval	Y: Approves the license with additional conditions
		N: Does not add conditions
	3. Reject	Y: Rejects the request
		N: Does not reject the request
	4. Refer to OLT	Y: Refers the decision to OLT
		N: Does not refer the decision to OLT
	5. Override OLT	Y: Overrides OLT decision
		N: Does not override OLT decision
Ontario Land Tribunal (OLT)	6. Approve	Y: Approves the application as part of an appeal
		N: Does not approve the application as part of an appeal
	7. Conditional Approval	Y: Approves the application with additional conditions as part of an appeal
		N: Does not approve the application with additional conditions
	8. Reject	Y: Rejects the request
		N: Does not reject the request
Ministry of Municipal Affairs and Housing (MMAH)	9. Minister's Zoning Order	Y: Issues an MZO
		N: Does not issue an MZO
James Dick Construction Limited (JDCL)	10. Proceed	Y: Proceeds with the application
		N: Withdraws the application
Teedon Pit Conflict		
CRH Canada Group Inc. (CRH)	1. Apply	Y: Applies for expansion N: Does not apply for expansion
	2. Bring the Case to OLT	Y: Appeals decision
		N: Does not appeal decision

(*continued*)

Table 1. (*continued*)

Decision Makers	Options	Explanation*
Local Government (LG)	3. Approve Amendment	Y: Approves relevant applications
		N: Does not approve relevant applications
Province	4. Approve License	Y: Issues the license for expansion
		N: Rejects the application for expansion
	5. Minister's Zoning Order	Y: Issues an MZO to re-zone land parcel
		N: Does not issue an MZO
	6. Approve with Condition	Y: Approves application with added conditions
		N: Does not approve application with added conditions
Opposition	7. Bring the Case to OLT	Y: Appeals decision
		N: Does not appeal decision
	8. Petition Federal Government	Y: Requests federal government intervention
		N: Does not request federal government intervention
Federal Government	9. Intervene	Y: Intervenes in decision
		N: Does not intervene in the conflict
Ontario Land Tribunal (OLT)	10 Approval	Y: Approves the application as part of an appeal
		N: Rejects the application as part of an appeal
	11. Condition	Y: Approves the application with conditions added
		N: Rejects the application as part of an appeal

*An option for a DM is a binary choice—an action that a DM may ("Y") or may not ("N") execute

other matters in Ontario. If the Provincial Government rejects the license application, the Applicant is again entitled to appeal [26]. For aggregate mining cases, appeals to Tribunal can come from any person or entity who submitted comments (both oral or written) at public meetings (i.e., Citizens) and from the Applicant. If Tribunal holds a hearing, it can direct the Provincial Government to issue an aggregate license, to issue

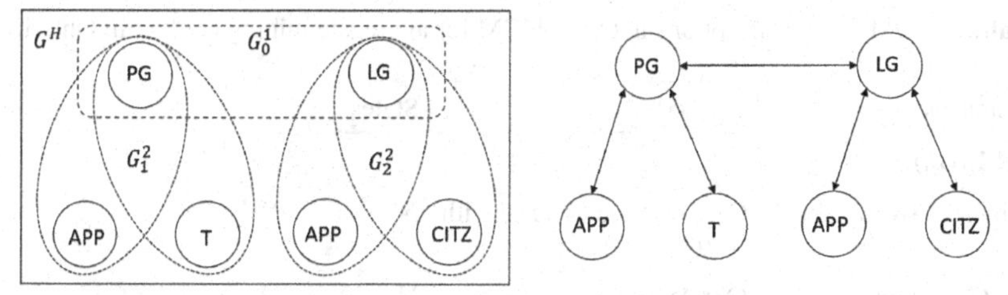

Fig. 3. Conceptual diagram for 2LHGM for aggregate mining conflict in Ontario (left) and interactions between DMs (right)

an aggregate license conditionally, or to refuse to issue a license. The province acts through provincial ministers. The MNDMNRF has the option to issue or refuse to issue an aggregate license. If the Tribunal directs to issue a license that includes new conditions, MNDMNRF can uphold those conditions or refuse to impose them [1]. The province can also intervene by initiating an MZO through the MMAH [29]. This power allows the MMAH to bypass local authorities and zone any property in the province.

3.7 Strategies, States, and Preference Relations in 2LHGM

Options in each conflict game correspond to the courses of action DMs have at their disposal at a given point in time. Each DM can choose to select (indicated by "Y") or not to select (indicated by "N") an option. A DM is said to have chosen a strategy when all options are selected or not. A conflict state is formed by all DMs choosing their strategies. To reduce model complexity, infeasible states—i.e., states that cannot happen in reality because they are mutually exclusive or are highly preferentially unlikely—are removed to generate a set of feasible conflict states. Preferences of each DM over the set of feasible state in a given conflict can be ranked using option prioritization [31, 35, 36], a process whereby statements about preferred states are used to sort the set of feasible states from most to least preferred from each DM's perspective. Preference statements can be expressed in logical terms, including negation ("NOT" or "−"), conjunction ("AND" or "&"), disjunction ("OR" or "|"), and conditions ("IF" and "IFF") [34].

Preferences in an HGM cannot be fully determined by the preference information in all sub-conflicts. Instead, different patterns of preference structure can be attained by sorting the constituent sub-conflicts in their order of importance [49]. Therefore, two types of preference structures can be defined within an HGM: one between the sub-models, which is from each CDM's perspective [48], and the other within each sub-model based on how each DM perceives the dispute. For the aggregate mining HGM, the former preferential pattern was not considered, as both the Provincial and Local Governments—the two CDMs of the study—have no preferences over the sub-models. Therefore, no weights are assigned for each sub-conflict. As for the within sub-model preferences, those for Citizens are shaped by their expectation that pits will negatively affect environmental quality; those for Applicant are determined based on their interest to get their applications approved; and those for other DMs (i.e., the Provincial Government, Local Government, and Tribunal) are shaped by their indifference over this class of conflicts in the province—this is mainly due to the different nature of each aggregate

Table 2. DMs and their options in the 2LHGM for aggregate mining conflict in Ontario.

Decision Maker	Options	States						
Level 1 (G_0^1)								
Provincial Government	O_1: Approve License with MZO	N	Y	N				
Local Government	O_2: Approve Land Use Amendment	N	N	Y				
State Number		1	2	3				
Level 2 (G_{11}^2)								
Provincial Government	O_3: Approve License	N	N	Y	N			
Applicant	O_4: Apply for a License	N	Y	Y	Y			
	O_5: Appeal	N	N	N	Y			
State Number		1	2	3	4			
Level 2 (G_{12}^2)								
Provincial Government	O_3: Approve License	N	N	Y	N	N	Y	N
	O_6: Approve License with Condition	N	N	N	Y	N	N	Y
Tribunal	O_7: Approve Appeal	N	Y	Y	Y	N	N	N
	O_8: Approve Appeal with Condition	N	N	N	N	Y	Y	Y
State Number		1	2	3	4	5	6	7
Level 2 (G_{21}^2)								
Local Government	O_2: Approve Land Use Amendment	N	N	Y	N			
Applicant	O_9: Apply for Land Use Amendment	N	Y	Y	Y			
	O_5: Appeal	N	N	N	Y			
State Number		1	2	3	4			
Level 2 (G_{22}^2)								
Local Government	O_2: Approve Land Use Amendment	N	Y	Y				
Citizens	O_{10}: Appeal	N	N	Y				
State Number		1	2	3				

*An option for a DM is a binary choice—an action that a DM may ("Y") or may not ("N") execute

mining conflict in the province; to generate a holistic model, it is imperative to assume an equal preference structure for these DMs across all the feasible states. Preference statements and rankings for the aggregate mining HGM are listed in Table 3. In this table, the preference statements are listed from the most important at the top to the least important at the bottom for each DM in each sub-model. Also, feasible states for each DM are listed from most preferred on the left to the least preferred on the right.

Table 3. State ranking and preference statements for each DM in different hierarchies.

Decision Maker	Decision Rule	State Ranking
Level 1 (G_0^1)		
Provincial Government	Indifferent	[1,2,3]
Local Government	Indifferent	[1,2,3]
Level 2 (G_{11}^2)		
Provincial Government	Indifferent	[1,2,3,4]
Applicant	(i) O_3	3,4,2,1
	(ii) NOT O_3 & O_5	
	(iii) O_4	
Level 2 (G_{12}^2)		
Provincial Government	Indifferent	[1,2,3,4,5,6,7]
Tribunal	Indifferent	[1,2,3,4,5,6,7]
Level 2 (G_{21}^2)		
Local Government	Indifferent	[1,2,3,4]
Applicant	(i) O_2	3,4,2,1
	(ii) NOT O_2 & O_5	
	(iii) O_9	
Level 2 (G_{22}^2)		
Local Government	Indifferent	[1,2,3]
Citizens	(i) NOT O_2	1,3,2
	(ii) O_2 & O_{10}	

*States enclosed in square brackets are equally preferred
*Oi are different options considered in defining the preferences. The corresponding option for each Oi is listed in Table 2

4 Results

In this section, an overview of the results is presented within each defined hierarchy of the aggregate mining conflict model. The Common patterns at each level are identified, and the overall 2LHGM results are presented.

4.1 Local Games

At the lowest hierarchy, the model yields the following results for the four local games considered in the analysis.

- G^2_{11}: Two states are in equilibrium in the game between PG and APP; these states occur when an application for a license is either approved or rejected by PG. Both cases are Nash and SEQ stable; the only difference is that APP appeals in the latter case.
- G^2_{12}: All states are Nash and SEQ stable in the game between PG and T.
- G^2_{21}: Two states are in equilibrium in the game between LG and APP; these states occur when an application for a change in zoning bylaw is either approved or rejected by LG. Both cases are Nash and SEQ stable; the only difference is that APP appeals in the latter case.
- G^2_{22}: Two states are in equilibrium in the game between LG and CITZ; these states occur when an application for a change in zoning bylaw is either approved or rejected by LG. Both cases are Nash and SEQ stable; the only difference is that CITZ appeals in the former case.

The potential outcomes attained from each of these local games present the common patterns about the prevailing non-cooperative circumstance of local aggregate games in the province. Assuming that PG, LG, and T remain indifferent in their preferences, the only outcomes resulting in appeal cases are when there is a conflict of interests between the APP or CITZ and the Governments. The results of all local games in the conflict are listed in Table 4.

4.2 Basic Hierarchical Graph Models

At the second hierarchy, G^2_1 includes five equilibria between the PG, APP, and T. All states are Nash and SEQ stable. All states present instances where PG does not approve an application, and APP appeals this decision. In G^2_2 there is no resolution for the game between LG, APP, and CITZ.

The added complexity represented by non-cooperative competing interests in G^2_1 and G^2_2 games at this level conforms to the status quo in the aggregate mining conflicts in the province: when APP and CITZ are playing the same game (i.e., G^2_2), no compromises are made, but when APP is at the same game with PG, APP, and T (G^2_1), all potential resolutions occur when there is an appeal on the part of APP. The results are listed in Table 5.

Table 4. Stability results of all local games in the conflict

G_0^1							
State	NN	YN	NY				
Stability (PG's View)	r,s	r,s	r,s				
Stability (LG's View)	r,s	r,s	r,s				
Equilibrium	E	E	E				
G_{11}^2							
State	NNN	NYN	YYN	NYY			
Stability (PG's View)	r,s	r,s	r,s	r,s			
Stability (APP's View)			r,s	r,s			
Equilibrium			E	E			
G_{12}^2							
State	NNNN	NNYN	YNYN	NYYN	NNNY	YNNY	NYNY
Stability (PG's View)	r,s	r,s	r,s	r,s	r,s	r,s	r,s
Stability (T's View)	r,s	r,s	r,s	r,s	r,s	r,s	r,s
Equilibrium	E	E	E	E	E	E	E
G_{21}^2							
State	NNN	NYN	YYN	NYY			
Stability (LG's View)	r,s	r,s	r,s	r,s			
Stability (APP's View)			r,s	r,s			
Equilibrium			E	E			
G_{22}^2							
State	NN	YN	YY				
Stability (LG's View)	r,s	r,s	r,s				
Stability (CITZ's View)	r,s		r,s				
Equilibrium	E		E				

*r, s, and E denote Nash stable, SEQ stable and equilibrium states, respectively

4.3 Two-Level Hierarchical Graph Model

Following the stability definitions presented for a 2LHGM in Sect. 3.3, none of the states for the overall hierarchical model for aggregate mining conflicts in Ontario (G^H) are in equilibrium. To determine the potential resolutions at this level, the results of G_1^2 and G_2^2 were combined with the results of G_0^1, a game between the two CDMs (i.e., PG and LG) in which all outcomes are in equilibria. The results for the G^H model are listed in Table 5.

Table 5. Stability result of two BHGMs and 2LHGM.

G_1^2			G_2^2	
State ($O_3O_4O_5O_6O_7O_8$)	N—NNNN	NYY———	State ($O_2O_5O_9O_{10}$)	NNNN
Stability (PG's View)	r,s	r,s	Stability (LG's View)	r,s
Stability (APP's View)		r,s	Stability (APP's View)	
Stability (T's View)	r,s	r,s	Stability (CITZ's View)	r,s
Equilibrium		E	Equilibrium	
G^H				
State ($O_1O_2O_3O_4O_5O_6O_7O_8O_9O_{10}$)			NNN—NNNNNN	
Stability (PG's View)			r,s	
Stability (LG's View)			r,s	
Stability (APP's View)				
Stability (T's View)			r,s	
Stability (CITZ's View)			r,s	
Equilibrium				

*—indicate that both "Y" and "N" options can be selected

5 Insights and Concluding Remarks

This study aimed to provide a more in-depth view of aggregate mining disputes in Ontario by constructing a generic 2LHGM for this class of provincial conflicts. The general structure of the proposed 2LHGM was determined based on three case-specific, one-level aggregate mining conflict models in the province [24–26]. In doing so, the authors focused on unveiling the structured hierarchies of power and strategic micro-level interactions in Ontario's aggregate mining disputes to shed light on potential outcomes (and possible resolutions) that might have been overlooked in the one-level conflict models of the past.

The results suggest that the considered preference structure for the CDMs—that of neutrality—does not create an equilibrium for this class of provincial conflicts. Therefore, potential resolutions at the highest level might be achieved when there is a detailed preference structure for the CDMs. Nevertheless, potential resolutions obtained from the BHGM that considers the Provincial Government as a CDM illustrate an indispensable role that this DM plays in the fate of the conflict at the lower hierarchy.

These findings corroborate the findings of past research on provincial aggregate management and planning system [18] by highlighting its limited capacity in finding compromised solutions between stakeholders in this class of provincial conflicts. Although the proposed conflict model differs from individual aggregate cases [24–26] in preference structure, it provides opportunities to examine the potential advocacy roles that different DMs could play in this class of conflicts by capturing the existing hierarchical relationships between them.

Future studies should therefore focus on exploring the potential impacts of changing these preference structures on conflict outcomes. This will be crucial to finding common ground between all DMs and reinvigorating public participation processes that have been debilitated due to the recent developments in Ontario's environmental policy-making direction [54].

References

1. Aggregate Resources Act (1990). https://www.ontario.ca/laws/statute/90a08. Accessed 27 Jan 2023
2. Drew, L.J., Langer, W.H., Sachs, J.S.: Environmentalism and natural aggregate mining. Nat. Resour. Res. **11**(1), 19–28 (2002). https://doi.org/10.1023/A:1014283519471
3. Langer, W.H., Arbogast, B.F.: Environmental impacts of mining natural aggregate. In: Fabbri, A.G., Gaál, G., McCammon, R.B. (eds.) Deposit and Geoenvironmental Models for Resource Exploitation and Environmental Security, pp. 151–169. Springer, Dordrecht (2002). https://doi.org/10.1007/978-94-010-0303-2_8
4. Wernstedt, K.: Plans, planners, and aggregates mining: constructing an understanding. J. Plan. Educ. Res. **20**(1), 77–87 (2000). https://doi.org/10.1177/073945600128992618
5. Poulin, R., Pakalnis, R.C., Sinding, K.: Aggregate resources: production and environmental constraints. Environ. Geol. **23**(3), 221–227 (1994). https://doi.org/10.1007/BF00771792
6. Bradshaw, A.: Restoration of mined lands—using natural processes. Ecol Eng **8**(4), 255–269 (1997). https://doi.org/10.1016/S0925-8574(97)00022-0
7. Hilson, G.: An overview of land use conflicts in mining communities. Land Use Policy **19**(1), 65–73 (2002). https://doi.org/10.1016/S0264-8377(01)00043-6
8. Dulias, R.: Landscape planning in areas of sand extraction in the silesian upland, Poland. Landsc. Urban Plan. **95**(3), 91–104 (2010). https://doi.org/10.1016/j.landurbplan.2009.12.006
9. Van Wagner, E.: Law's rurality: land use law and the shaping of people-place relations in rural Ontario. J. Rural Stud. **47**, 311–325 (2016). https://doi.org/10.1016/j.jrurstud.2016.01.006
10. Schiappacasse, P., Müller, B., Linh, L.T.: Towards responsible aggregate mining in Vietnam. Resources **8**(3), 138 (2019). https://doi.org/10.3390/resources8030138
11. Esteves, A.M.: Mining and social development: refocusing community investment using multi-criteria decision analysis. Resour. Policy **33**(1), 39–47 (2008). https://doi.org/10.1016/j.resourpol.2008.01.002
12. Ontario Aggregate Resources Corporation: Aggregate Resources Statistics in Ontario (2019). https://toarc.com/wp-content/uploads/2021/02/Stats_2019_Final.pdf. Accessed 27 Jan 2023
13. Binstock, M., Carter-Whitney, M.: Aggregate Extraction in Ontario: A Strategy for the Future (2011). http://cielap.org/pdf/AggregatesStrategyExecSumm.pdf. Accessed 27 Jan 2023
14. Statistics Canada: Table 17-10-0009-01 Population estimates, quarterly (2021). https://www150.statcan.gc.ca/t1/tbl1/en/tv.action?pid=1710013901
15. Statistics Canada: Table 36-10-0487-01 Gross domestic product (GDP) at basic prices, by sector and industry, provincial and territorial (x 1,000,000) (2021). https://www150.statcan.gc.ca/t1/tbl1/en/tv.action?pid=3610048701
16. Winfield, M.S., Taylor, A.: Rebalancing the Load: The need for an aggregates conservation strategy for Ontario (2005). https://www.pembina.org/reports/Aggregatesfinal-web2.pdf. Accessed 27 Jan 2023
17. Kellett, J.E.: The elements of a sustainable aggregates policy. J. Environ. Planning Manage. **38**(4), 569–580 (1995). https://doi.org/10.1080/09640569512832

18. Port, C.M.: The Opportunities and Challenges of Aggregate Site Rehabilitation in Southern Ontario. An Evaluation of the Rehabilitation Process from 1992–2011 (2013). http://hdl.handle.net/10012/7966
19. Baker, D., Shoemaker, D.: Environmental Assessment and Aggregate Extraction in Southern Ontario: The Puslinch Case (1995). https://uwaterloo.ca/applied-sustainability-projects/sites/default/files/uploads/documents/ontario_3_bakershoemaker_puslinch_aggregates.pdf. Accessed 27 Jan 2023
20. Markvart, T.I.: Understanding Institutional Change and Resistance to Change Towards Sustainability: An Interdisciplinary Theoretical Framework and Illustrative Application to Provincial-Municipal Aggregates Policy (2009). http://hdl.handle.net/10012/4653
21. Chambers, C., Sandberg, A.L.: Pits, peripheralization and the politics of scale: struggles over locating extractive industries in the town of Caledon, Ontario, Canada. Reg. Stud. **41**(3), 327–338 (2007). https://doi.org/10.1080/00343400600928319
22. Patano, S., Sandberg, L.A.: Winning back more than words? Power, discourse and quarrying on the niagara escarpment. Canadian Geographies/Les géographies canadiennes **49**(1), 25–41 (2005). https://doi.org/10.1111/j.0008-3658.2005.00078.x
23. Baker, D., Slam, C., Summerville, T.: An evolving policy network in action: the case of construction aggregate policy in Ontario. Can. Public Adm. **44**(4), 463–483 (2001). https://doi.org/10.1111/j.1754-7121.2001.tb00901.x
24. Philpot, S., Hipel, K.W.: We are going to make sure it doesn't happen one way or another'. investigating a proposed quarry in Canada. In: Chang, N.-B., Fang, L. (eds.) Proceeding of 9th International Conference on Water Resources and Environment Research (ICWRER), pp. 21–26. University of Central Florida, Orlando (2022)
25. Philpot, S., Mirnasl, N., Hipel, K.W.: Conflict in tiny town: aggregate mining at the alliston aquifer. In: Morais, D.C., Fang, L. (eds.) Group Decision and Negotiation: Methodological and Practical Issues, pp. 74–90. Springer, Cham (2022). https://doi.org/10.1007/978-3-031-07996-2_6
26. Philpot, S., Hipel, K.W.: Investigating an aggregate mine proposal using the graph model for conflict resolution. Ann. Am. Assoc. Geogr. **112**(6), 1812–1832 (2022). https://doi.org/10.1080/24694452.2021.1994850
27. Philpot, S., Johnson, P.A., Hipel, K.W.: Analysis of a below-water aggregate mining case study in Ontario, Canada using values-centric online citizen participation. J. Environ. Planning Manage. **63**(2), 352–368 (2020). https://doi.org/10.1080/09640568.2019.1588713
28. Government of Ontario: Aggregate resources (2014). https://www.ontario.ca/page/aggregate-resources. Accessed 27 Jan 2023
29. Planning Act (1990). https://www.ontario.ca/laws/statute/90p13
30. Ministry of Northern Development Mines Natural Resources and Forestry: Aggregate resources (2021). https://www.ontario.ca/page/aggregate-resources. Accessed 30 Jan 2023
31. Xu, H., Hipel, K.W., Kilgour, D.M., Fang, L.: Conflict Resolution Using the Graph Model: Strategic Interactions in Competition and Cooperation. Springer, Cham (2018). https://doi.org/10.1007/978-3-319-77670-5
32. Hipel, K.W., Fang, L.: The graph model for conflict resolution and decision support. IEEE Trans. Syst. Man Cybern. Syst. **51**(1), 131–141 (2021). https://doi.org/10.1109/TSMC.2020.3041462
33. Kilgour, D.M., Hipel, K.W.: The graph model for conflict resolution: past, present, and future. Group Decis. Negot. **14**(6), 441–460 (2005). https://doi.org/10.1007/s10726-005-9002-x
34. Fang, L., Hipel, K.W., Kilgour, D.M., Peng, X.: A decision support system for interactive decision making-Part I: model formulation. IEEE Trans. Syst. Man Cybern. Part C (Appl. Rev.) **33**(1), 42–55 (2003). https://doi.org/10.1109/TSMCC.2003.809361

35. Hipel, K.W., Fang, L., Kilgour, D.M.: The graph model for conflict resolution: reflections on three decades of development. Group Decis. Negot. **29**(1), 11–60 (2019). https://doi.org/10.1007/s10726-019-09648-z
36. Fang, L., Hipel, K.W., Kilgour, D.M.: Interactive Decision Making: The Graph Model for Conflict Resolution. Wiley (1993)
37. Nash, J.: Equilibrium points in N-person games. Proc. Natl. Acad. Sci. **36**(1), 48–49 (1950). https://doi.org/10.1073/pnas.36.1.48
38. Nash, J.: Non-cooperative games. Ann. Math. **54**(2), 286–295 (1951). https://doi.org/10.2307/1969529
39. Fraser, N., Hipel, K.W.: Conflict Analysis: Models and Resolutions. North-Holland, Amsterdam (1984)
40. Howard, N.: Paradoxes of Rationality Theory of Metagames and Political Behavior. MIT Press, Cambridge (1971)
41. Kilgour, D.M., Hipel, K.W.: Conflict analysis methods: the graph model for conflict resolution. In: Kilgour, D.M., Eden, C. (eds.) Handbook of Group Decision and Negotiation, pp. 203–222. Springer, Dordrecht (2010). https://doi.org/10.1007/978-90-481-9097-3_13
42. Hämäläinen, R.P., Luoma, J., Saarinen, E.: On the importance of behavioral operational research: the case of understanding and communicating about dynamic systems. Eur. J. Oper. Res. **228**(3), 623–634 (2013). https://doi.org/10.1016/j.ejor.2013.02.001
43. Mingers, J., White, L.: A review of the recent contribution of systems thinking to operational research and management science. Eur. J. Oper. Res. **207**(3), 1147–1161 (2010). https://doi.org/10.1016/j.ejor.2009.12.019
44. Beimel, A., Tassa, T., Weinreb, E.: Characterizing Ideal weighted threshold secret sharing. In: Kilian, J. (ed.) TCC 2005. LNCS, vol. 3378, pp. 600–619. Springer, Heidelberg (2005). https://doi.org/10.1007/978-3-540-30576-7_32
45. Pratt, T.W.: Definition of programming language semantics using grammars for hierarchical graphs. In: Claus, V., Ehrig, H., Rozenberg, G. (eds.) Graph-Grammars and Their Application to Computer Science and Biology, pp. 389–400. Springer, Heidelberg (1979). https://doi.org/10.1007/BFb0025735
46. Busatto, G., Kreowski, H., Kuske, S.: Abstract hierarchical graph transformation. Math. Struct. Comput. Sci. **15**(4), 773–819 (2005). https://doi.org/10.1017/S0960129505004846
47. Drewes, F., Hoffmann, B., Plump, D.: Hierarchical graph transformation. J. Comput. Syst. Sci. **64**(2), 249–283 (2002). https://doi.org/10.1006/jcss.2001.1790
48. He, S., Kilgour, D.M., Hipel, K.W., Bashar, M.A.: A basic hierarchical graph model for conflict resolution with application to water diversion conflicts in China. INFOR Inf. Syst. Oper. Res. **51**(3), 103–119 (2013). https://doi.org/10.3138/infor.51.3.103
49. He, S.: Hierarchical Graph Models for Conflict Resolution (2015). http://hdl.handle.net/10012/9826
50. He, S., Hipel, K.W., Xu, H., Chen, Y.: A two-level hierarchical graph model for conflict resolution with application to international climate change negotiations. J. Syst. Sci. Syst. Eng. **29**(3), 251–272 (2020). https://doi.org/10.1007/s11518-019-5448-2
51. He, S., Marc Kilgour, D., Hipel, K.W.: A general hierarchical graph model for conflict resolution with application to greenhouse gas emission disputes between USA and China. Eur. J. Oper. Res. **257**(3), 919–932 (2017). https://doi.org/10.1016/j.ejor.2016.08.014
52. IBI Group: Planning summary report proposed Hallman Pit 1894 Witmer Road, Wilmot Twp (2019). https://facility-admin.esolutionsgroup.ca/Uploads/Files/16E7D05A-FC42-4E34-A1EF-8C5C6858A2BF/zca-11-19/Planning%20Summary%20Report.pdf. Accessed 27 Jan 2023
53. Government of Ontario: Environmental Registry of Ontario (ERO) (2021). https://ero.ontario.ca/. Accessed 30 Nov 2020

54. Mirnasl, N., Philpot, S., Akbari, A., Hipel, K.W.: Assessing policy robustness under the COVID-19 crisis: an empirical study of the environmental policymaking system in Ontario, Canada. J. Environ. Planning Policy Manage. **24**(6), 762–776 (2022). https://doi.org/10.1080/1523908X.2022.2051454

Weighted Reward Allocation Mechanism for Data Collection Quality

Ayato Kitadai(✉), Sinndy Dayana Rico Lugo, Sangjic Lee, Masanori Fujita, and Nariaki Nishino

The University of Tokyo, Tokyo 113-8656, Japan
a.kitadai@css.t.u-tokyo.ac.jp
https://tmi.t.u-tokyo.ac.jp/

Abstract. Data utilization, which offers several benefits in real-world contexts, has caught the attention of several researchers and organizations. However, obtaining high-quality data is difficult. Moreover, it usually requires a large amount of compensation when the data are intended for private use. Recently, the development and refinement of methods to incentivize people to provide useful data, called mechanisms, has received special attention, consequently increasing demand for more efficient models. This study was conducted to propose a mechanism, "Weighted Reward Allocation Mechanism (WRAM)", for promoting high-quality data supplied by rational data submitters considering budget constraints. The mechanism calculates and distributes the total budget in accordance with a weighted contribution evaluation vector of each item in a questionnaire. After mathematical details of WRAM are introduced, a simulation experiment is presented by which the data quality is compared with that of a widely used uniform reward mechanism. Results demonstrated that the submitter behavior changes depending on WRAM weights. Overall, our proposed mechanism performs better than the uniform reward mechanism in terms of the quality of the collected data.

Keywords: Reward mechanism · Data collection · Game theory

1 Introduction

In recent years, data utilization has attracted a great deal of attention in real-world contexts such as healthcare and retail, government services, and manufacturing and location-based services. Benefits, collection and analysis methods, and resources necessary for data use are varied [10]. Therefore, the use of machine learning for data utilization has increased within many research and business fields, necessitating the development of learning algorithms that are more accurate and efficient.

Furthermore, high-quality and large datasets are necessary to make use of data. Nevertheless, preparing them independently is expensive, leading many researchers to use open datasets. Unfortunately, datasets of several types are

Supported by Otsuka Toshimi Scholarship Foundation.

Y. Maemura et al. (Eds.): GDN 2023, LNBIP 478, pp. 161–174, 2023.
https://doi.org/10.1007/978-3-031-33780-2_11

not necessarily open to the public. It is particularly true that the quantitative limits of existing datasets have become apparent and that the lack of training data come to pose a major difficulty. Resolving those difficulties is expected to require the collection of new data and the creation of new datasets.

However, collecting new data is not easy. For instance, data collection from individuals, such as through questionnaires, requires the cooperative efforts of numerous people, but few people are willing to contribute free of charge. Consequently, paying a reward for cooperation is commonly done. Even then, however, cooperation is not ensured. Even if cooperation is obtained, then questionnaires might not be answered seriously. As a result, difficulties often arise such as the poor quality of the collected data.

Therefore, this study was conducted to propose a mechanism for promoting high-quality data supply by rational data submitters under a fixed reward budget, called the "Weighted Reward Allocation Mechanism (WRAM)". Questionnaires with multiple-choice responses and descriptive questions were used to analyze data collection incentives for private use. A new theoretical solution was proposed and then compared with the commonly used Nash equilibrium. Consequently, this study is the first attempt to provide a basis for smoother reward allocation mechanisms intended for increasing the quality of data collected through data-sharing schemes. The remainder of this article includes the summary of related literature (Sect. 2), explanation of details of our proposed model (Sect. 3), description of a simulation experiment (Sect. 4), and discussion of the relevant findings and conclusions (Sect. 5).

2 Related Literature

Several studies have defined data quality depending on the context [1,6,7,20], usually referring to data reliability and the fulfillment of user requirements [9]. Redman [15,16] and Strong et al. [19] define high-quality data as fit for purpose, whereas Wang and Strong [20] defined data quality in terms of data usability.

In general, if low-quality data are used, then companies end up losing money [14] or producing defective products [4]. Consequently, quality data can be understood as having comprehensive and easily accessible information for making evidence-based decisions, including information related to best practices and customer preferences [2,18]. Especially, several concerns have arisen about the use of respondents who are rewarded irrespective of the response contents because of their potential intrinsic motivation to complete a questionnaire carelessly, merely to receive the payment [5,17]. This practice impairs the accuracy and completeness of the collected data, leading to eventual data quality deterioration.

Considering that game theory mathematical models are intended to represent people's decision-making and to facilitate analyses of an equilibrium state for improving social conditions, researchers have used them to elucidate mechanisms necessary to obtain useful responses during data collection. Earlier research within this field includes the Peer-Prediction Method [11], which uses a scoring system based on the stochastic correlation between the reports of different raters, in combination with appropriate rewards, to incentivize useful reporting. The key idea of the Peer-Prediction Method is the comparison of the inferred probabilities with the report of a benchmark rater instead of a comparison of actual reports.

The Bayesian Truth Serum (BTS) [13] is another mechanism for collecting people's honest responses. It rewards high scores to answers that are more frequent than what would be predicted by the same population, rather than just the most common answers. It maximizes the expected score even for respondents who believe their answer is a minority opinion. Nevertheless, this method is limited to use with multiple-choice questions [12].

In this context, Kyeong and Nam [8] introduced a new theoretical model to improve the reliability of data used for business intelligence. That mechanism incorporates the concept of influence through a network-based reasoning model. Using the two-step flow theory, they found that using influence-based data collection methods engenders data collection and analysis that are more efficient.

3 Model for Data Collection

Within the context of data collection, there are private and public uses. In the former, the purpose of the utilization does not benefit the submitter, whereas the latter does. In this study, we theoretically analyzed how rational the decision-making process of the data submitter will be in the context of private use after modeling the data collection from a game theory perspective.

Let $N = \{1, 2, \ldots, n\}$ be the set of data submitters (players). Also, let the data owned by any player $j \in N$ be $\boldsymbol{d}^j = (d_1^j, d_2^j, \ldots, d_k^j)$. Therein, d_i^j represents the concrete content of data item i owned by player j. We envision a large data collection of thousands of people and assume that the number of players n is sufficiently large.

Any player $j \in N$ bears the provision cost c_i^j when providing each datum d_i^j. We designate the combination $(c_1^j, c_2^j, \ldots, c_k^j)$ as the type of player j. This designation reflects the effort necessary to arrange and provide each data item and the willingness to provide it, which differ among players.

Here, each player knows the player's own type, i.e., the set of provision costs that the player has, but does not know the types of the other players. However, players know that for $\forall i \in \{1, 2, \ldots, k\}$, the provision cost c_i^j of the data d_i^j held by any player $j \in N$ follows an independently continuous distribution function F_i.

The data collector gives all players a reward budget R plus the player's strategy set S_j. Then, each player $j \in N$ conveys the distribution rule of the reward r_j received according to the combination of his own strategy s_j and the strategies of the other players s_{-j}.

Let the player's strategy $s_j \in S_j$ be a subset of $\{1, 2, \ldots, k\} \equiv s$, which represents the data items honestly provided by player j. For example, $s_j = \{1, 4\}$ corresponds to player j providing data d_1^j, d_4^j but no other data, whereas $s_j = \phi$

corresponds to player j providing no data at all. In this paper, we assume that all players do not lie and honestly provide accurate data they own.

Using reward r_j and the provided cost c_i^j, the player's utility function u_j is defined as follows.

$$u_j(s_j, s_{-j}) := r_j(s_j, s_{-j}) - \sum_{i \in s_j} c_i^j. \tag{1}$$

Subsequently, data collection from individuals for private use is modeled as a Bayesian game. All players rationally make decisions to maximize their utility. Additionally, we considered that the data collector wants to use a mechanism in which the quality of the collected data q, determined by the Bayesian Nash equilibrium of the game, reaches the threshold t under budget constraints.

3.1 Uniform Reward Mechanism

For general data collection, a uniform reward mechanism is adopted, giving a uniform reward to all data submitters. For discussion in this section, we formulated this mechanism as one in which players can choose to either provide all or none of their data: an arbitrary player $j \in N$'s strategy set is $S_j = \{s, \phi\}$. Each player has a strategy $s_j \in \{s, \phi\}$. Consequently, $s_j = s$ corresponds to the provision of all data. In addition, $s_j = \phi$ corresponds to not providing any data at all.

Here, the reward r_j for each player $j \in N$ is independent of the strategy combination s_{-j} of the other players, i.e., a predetermined amount is given only when data are provided ($s_j = s$). Because the total reward $\sum_{j=1}^{n} r_j$ to the player must not exceed the reward budget R, r_j is defined as follows:

$$r_j(s_j, s_{-j}) = \begin{cases} R/n & (\text{if } s_j = s), \\ 0 & (\text{if } s_j = \phi). \end{cases} \tag{2}$$

From (1) and (2), the utility function u_j for player j is

$$\begin{cases} u_j(s) = R/n - \sum\limits_{i=1}^{k} c_i^j, \\ u_j(\phi) = 0 - \sum\limits_{i \in \phi} c_i^j = 0. \end{cases} \tag{3}$$

Therefore, a rational player j chooses a strategy that maximizes the player's own utility function (3). Then, the Bayesian Nash equilibrium is that all players formulate the strategy s_j according to the action strategy (4).

$$s_j = \begin{cases} s \ (\text{if } R/n \geq \sum\limits_{i=1}^{k} c_i^j), \\ \phi \ (\text{otherwise}). \end{cases} \tag{4}$$

3.2 Weighted Reward Allocation Mechanism

In our proposed mechanism, called the Weighted Reward Allocation Mechanism (WRAM), the contribution from each player's strategy is calculated using the value of the weighted contribution evaluation vector $\boldsymbol{w} = (w_1, w_2, \ldots, w_k)$, which the data collector defines. It is a mechanism that computes and distributes the total reward budget R according to that value. For simplicity, the weighted evaluation vector is simply designated as the weight.

First, the data collector determines the value of the weight[1] $\boldsymbol{w}$ to be $\sum_{i=1}^{k} w_i = 1$. Here, the player's strategy set is $S_j := \{m \mid m \subseteq \{1, 2, \ldots, k\}\}$, which means that all players can freely choose any combination of the data items they provide.

The contribution C_j of player j is defined as the sum of the weights of the data items provided by j as

$$C_j(s_j) := \sum_{i \in s_j} w_i \in [0, 1]. \tag{5}$$

Subsequently, the reward r_j received by player j is

$$r_j(s_j, s_{-j}) := \frac{RC_j}{\sum_{i \in N} C_i}. \tag{6}$$

Next, to derive the Bayesian Nash equilibrium under WRAM, we computed the utility function of player j. By setting $\sum\limits_{j' \in N \setminus \{j\}} C_{j'} = C_{-j} \in [0, n-1]$, we obtained

$$r_j = \frac{RC_j}{C_{-j} + C_j}.$$

Here, we assume that n is sufficiently large and a player's contribution is much smaller than the sum of all other players' contributions. Therefore, $C_{-j} = \sum\limits_{j' \in N \setminus \{j\}} C_{j'} \gg C_j$. Then,

$$r_j = \frac{RC_j}{C_{-j}}. \tag{7}$$

Using this, the utility function for player j is

$$u_j = \frac{RC_j}{C_{-j}} - \sum_{i \in s_j} c_i^j = \sum_{i \in s_j} \left(\frac{Rw_i}{C_{-j}} - c_i^j \right). \tag{8}$$

From (8), the expected utility of player j is

$$\mathbb{E}(u_j) = \sum_{i \in s_j} \left(\mathbb{E}\left(\frac{1}{C_{-j}} \right) Rw_i - c_i^j \right). \tag{9}$$

---

[1] Here, weight $\boldsymbol{w}$ is simply a coefficient defined to determine the contribution of each player. It is determined independently of the importance of each data item.

Furthermore, conventional game theory approaches assume that players make decisions to maximize the expected payoff determined by (9). However, numerical analysis based on expected utility is very difficult to be conducted at this time. Therefore, for this study, we analyzed this mechanism by taking a different approach from the conventional game theory.

As presented earlier (8), C_{-j} is a value determined by other players. Also, R, w_i, c_i^j are given constants. Subsequently, each player calculates the expected value $\mathbb{E}(C_{-j})$ of the total contribution C_{-j} of other players and makes decisions based on that calculation. In addition, all players have expected utility[2] that shall be maximized.

$$\tilde{\mathbb{E}}(s_j) = \sum_{i \in s_j} \left(\frac{Rw_i}{\mathbb{E}(C_{-j})} - c_i^j \right) \tag{10}$$

Because this game is a symmetric game, if $\mathbb{E}(C_{-j})$ can be calculated, its value should not depend on j. Consequently, if $\mathbb{E}(C_{-j}) = \mu \in [0, n-1]$, then

$$\tilde{\mathbb{E}}(u_j) = \sum_{i \in s_j} \left(\frac{Rw_i}{\mu} - c_i^j \right). \tag{11}$$

Also, the behavioral strategy which maximizes the expected utility of any player j is expressed as follows.

$$\begin{cases} i \in s_j & (\text{if } \frac{Rw_i}{\mu} > c_i^j) \\ i \notin s_j & (\text{otherwise}) \end{cases} \tag{12}$$

Below, we found that $\mu = \mathbb{E}(C_{-j})$ when all players follow the action strategy (12).

$$w_i^j = \begin{cases} w_i & (\text{if } \frac{Rw_i}{\mu} > c_i^j) \\ 0 & (\text{otherwise}) \end{cases} \tag{13}$$

By introducing a variable (13), the contribution of player j can be expressed for $\forall j \in N, \forall i \in \{1, \ldots, k\}$ as

$$C_j = \sum_{i \in s_j} w_i = \sum_{i=1}^{k} w_i^j. \tag{14}$$

Moreover, because the probability $\frac{Rw_i}{\mu} > c_i^j$ becomes $F_i(\frac{Rw_i}{\mu})$, then

$$\mathbb{E}(w_i^j) = F_i\left(\frac{Rw_i}{\mu}\right) w_i.$$

Therefore, the expected value of contribution C_j for any player $j \in N$ is

$$\mathbb{E}(C_j) = \sum_{i=1}^{k} F_i\left(\frac{Rw_i}{\mu}\right) w_i \tag{15}$$

[2] In conventional game theory, expected utility is used to mean the expected value of utility. However, the expected utility $\tilde{\mathbb{E}}(u_j)$ here does not represent the mathematical expected value of utility, but the expected utility based on the expected value.

Because $\mathbb{E}(C_j)$ is independent of j, it was obtained that

$$\mu = (n-1)\sum_{i=1}^{k} F_i\Big(\frac{Rw_i}{\mu}\Big)w_i \quad (\because (15)) \tag{16}$$

Here, equation (16) has a unique solution: $\mu^* \in [0, n-1]$.

Proof. Given that the distribution function F_i is a monotonically increasing continuous function for any i, $F_i(\frac{Rw_i}{x})$ is a monotonically decreasing continuous function for x. Therefore, the function (17) is also a monotonically decreasing continuous function for x.

$$f(x) := (n-1)\sum_{i=1}^{k} F_i\Big(\frac{Rw_i}{x}\Big)w_i \tag{17}$$

In addition,

$$\lim_{x\to+0} f(x) = (n-1)\sum_{i=1}^{k}\Big(w_i \lim_{t\to\infty} F_i(t)\Big) = n-1. \tag{18}$$

Therefore, line $y = x$ and curve $y = f(x)$ have only one intersection point at (μ^*, μ^*) and $0 < \mu^* < n-1$. □

From the proof presented above, the optimal response for player j is the Bayesian Nash equilibrium, where all players choose the action strategy (19).

$$\begin{cases} i \in s_j & (\text{if } \frac{Rw_i}{\mu^*} > c_i^j) \\ i \notin s_j & (\text{otherwise}) \end{cases} \tag{19}$$

4 Simulation Experiment

4.1 Description

For an arbitrary data item d_i, when the provision cost follows a uniform distribution $F_i = [\underline{c_i}, \overline{c_i}]$, the behavior strategy in the equilibrium state, μ^* in (19), is calculable as the only solution of (16) as

$$\mu = (n-1)\sum_{i=1}^{k} F_i\Big(\frac{Rw_i}{\mu}\Big)w_i$$

Because F_i is a uniform distribution on the closed interval $[\underline{c_i}, \overline{c_i}]$, we obtain the following:

$$F_i\Big(\frac{Rw_i}{\mu}\Big) = \begin{cases} 1 & (\text{if } \frac{Rw_i}{\mu} \geq \overline{c_i}) \\ \frac{Rw_i/\mu - \underline{c_i}}{\overline{c_i} - \underline{c_i}} & (\text{if } \underline{c_i} \leq \frac{Rw_i}{\mu} \leq \overline{c_i}) \\ 0 & (\text{if } \frac{Rw_i}{\mu} \leq \underline{c_i}) \end{cases} \tag{20}$$

Therein, if $\forall i,\ \underline{c_i} \le \frac{Rw_i}{\mu} \le \overline{c_i}$, then

$$\forall i,\ F_i\Big(\frac{Rw_i}{\mu}\Big) = \frac{Rw_i/\mu - \underline{c_i}}{\overline{c_i} - \underline{c_i}}. \tag{21}$$

Consequently, μ^* in the action strategy (19) is calculable as

$$\mu^* = \frac{n-1}{2}\Big(- b + \sqrt{b^2 + \frac{4aR}{n-1}}\Big).\quad (\because (16), (21)) \tag{22}$$

However, for simplicity, we defined a and b as

$$a = \sum_{i=1}^{k} \frac{w_i^2}{\overline{c_i} - \underline{c_i}},\ b = \sum_{i=1}^{k} \frac{\underline{c_i} w_i}{\overline{c_i} - \underline{c_i}}.$$

This solution is derived under the assumption that $\underline{c_i} \le \frac{Rw_i}{\mu} \le \overline{c_i}$ holds for any i. Although its validity remains unclear, it is analytically difficult to find a solution without it. Therefore, we assumed that the value of (22) is sufficiently close to the exact solution. Using this value, WRAM is optimized with the equilibrium state in which all players adopt the action strategy (19).

Moreover, considering that data collectors want to use a mechanism that allows the quality of the collected data $q \in \mathbb{R}$ to reach the predetermined target value t, we used q as the evaluation index of the mechanism.

The evaluation vector of the data item of the collected data, which reflects its relative importance in utilization, is $\boldsymbol{v} = (v_1,\ v_2,\ \ldots,\ v_k) \in \mathbb{R}^k$, satisfying $\sum_{i=1}^{m} v_i = 1$. For simplicity, the evaluation vector $\boldsymbol{v}$ of the data item is simply called the evaluation value. Also, v_i is called the evaluation value of the data item i. Both $\boldsymbol{v}$ and v_i are constants given by the data collector in advance, irrespective of the reward budget R and the threshold t, and different from the value of a data item for the data submitters w_k. Subsequently, the quality of the collected data q is defined as the amount of usefulness the data collector derives from the data as

$$q = \sum_{j \in N} \sum_{i \in s_j} v_i. \tag{23}$$

This is the amount of evaluation value of the data item the data collector receives as the consequence of deciding the optimal values of the weight $\boldsymbol{w}$ under the given values of the data importance for himself $\boldsymbol{v}$.

Under the uniform reward mechanism, the player's strategy set in the equilibrium state is uniquely determined by (4) after the player's type is determined. Furthermore, the player's strategy set is $\{s, \phi\}$ and $\sum\limits_{i \in s} v_i = 1,\ \sum\limits_{i \in \phi} v_i = 0$. Then, the quality of the collected data is represented by the expressions below.

$$\begin{aligned} q = \sum_{j \in N} \sum_{i \in s_j} v_i &= \sum_{j \in \{j' \in N \mid s_{j'} = \phi\}} \sum_{i \in \phi} v_i + \sum_{j \in \{j' \in N \mid s_{j'} = s\}} \sum_{i \in s} v_i \\ &= \#\{j \in N \mid s_j = s\} =: q_u \end{aligned} \tag{24}$$

On the other hand, under WRAM, after the player's type is determined by nature, its strategy set $s^* = (s_1^*, s_2^*, \dots, s_n^*)$ in the Bayesian Nash equilibrium is determined by the value of the weight $\boldsymbol{w} \in \mathbb{R}^k$ from (19). Here, the mapping $\tau_j : \mathbb{R}^k \to S_j$ represents the correspondence between the weight $\boldsymbol{w} \in \mathbb{R}^k$ and strategy $s_j^* \in S_j$ in the equilibrium state of any player j. Therefore, the quality of collected data q depends on the weight $\boldsymbol{w}$ set by the data collector as

$$q = \sum_{j \in N} \sum_{i \in \tau_j(\boldsymbol{w})} v_i =: G(\boldsymbol{w}). \tag{25}$$

For clarification, the data quality under the uniform reward mechanism is expressed separately from q_u. Under WRAM, it is expressed separately from $G : \mathbb{R}^k \to \mathbb{R}$ using the function $G(\boldsymbol{w})$.

Next, assuming a crowd-sourced questionnaire (outsourcing of tasks to a large group of people through an online network), we set the distribution of the provision cost F_i and the player's actual provision cost c_i^j. Then, we calculated the quality of collected data q under each mechanism. Using as reference the real-world data set of market prices for questionnaire surveys of CrowdWorks [3], which is the real-world crowdsourcing service in Japan and uses the uniform reward mechanism, we defined the settings of our hypothetical questionnaire. It had $n = 100$ as the number of players (survey respondents), and $k = 5$ as the number of survey items. It is assumed that questionnaire items d_1 and d_2 are multiple-choice questions, d_3, d_4, and d_5 are descriptive questions, and that the player's submission cost for each question had the distribution $F_1, F_2 : [5, 15]$, $F_3, F_4 : [40, 140]$, and $F_5 : [100, 300]$. Additionally, we considered four scenarios for the $\boldsymbol{v}$ of each data item as shown below.

- All scores are equal to the expected provision cost ratio (ASEC): $\boldsymbol{v}_1 = (0.025, 0.025, 0.225, 0.225, 0.5)$
- Low scores are for multiple-choice items and high scores are for descriptive items (LMHD): $\boldsymbol{v}_2 = (0.05, 0.05, 0.3, 0.3, 0.3)$
- Uniform scores (US): $\boldsymbol{v}_3 = (0.2, 0.2, 0.2, 0.2, 0.2)$
- High scores are for multiple-choice items and low scores are for descriptive items (HMLD) $\boldsymbol{v}_4 = (0.44, 0.44, 0.04, 0.04, 0.04)$

Therefore, first, for arbitrary player $j \in N$ in the simulation, c_i^j is determined according to the probability distribution F_i. Each player's type $(c_1^j, c_2^j, \dots, c_k^j)$ is fixed. Then, for each reward budget R and evaluation value $\boldsymbol{v}$, the quality of collected data q when using the uniform reward mechanism was calculated based on (4). In addition, under the same player type, the quality of collected data $G(\boldsymbol{w})$ when using WRAM for each R and evaluation value $\boldsymbol{v}$ was calculated based on (19).

As the weight $\boldsymbol{w}$ of each data item, we considered all patterns $(0, 0, 0, 0, 1), (0, 0, 0, 0.01, 0.99), \dots, (1, 0, 0, 0, 0)$ that are the sum $\sum_{i=1}^{5} w_i = 1$ in increments of 0.01. Then we ascertained the maximum value $\max G(\boldsymbol{w})$ and the minimum value $\min G(\boldsymbol{w})$ of the quality of the collected data.

4.2 Results

Figures 1a–1b show calculation results when the evaluated values $\boldsymbol{v} = \boldsymbol{v}_1, \boldsymbol{v}_2, \boldsymbol{v}_3, \boldsymbol{v}_4$. The horizontal axis represents the budget for compensation R. The vertical axis represents q. Solid and dotted lines respectively represent $\max G(\boldsymbol{w})$ and $\min G(\boldsymbol{w})$ at each R. The dash-dot line shows q_u when using the uniform reward mechanism as a reference for comparison between the mechanisms, which is the same for all figures because and independent of the value of $\boldsymbol{v}$. Additionally, Fig. 2 presents the result of calculating the value of $G(\boldsymbol{w}^*)$ under WRAM with $\boldsymbol{w}$ as $\boldsymbol{w}^* = (0.025, 0.025, 0.225, 0.225, 0.5)$ in the same ratio as the expected value of the provision cost. As in Fig. 1, the dash-dot and solid lines represent q and $\max G(\boldsymbol{w})$ respectively, whereas the dashed line represents $G(\boldsymbol{w}^*)$.

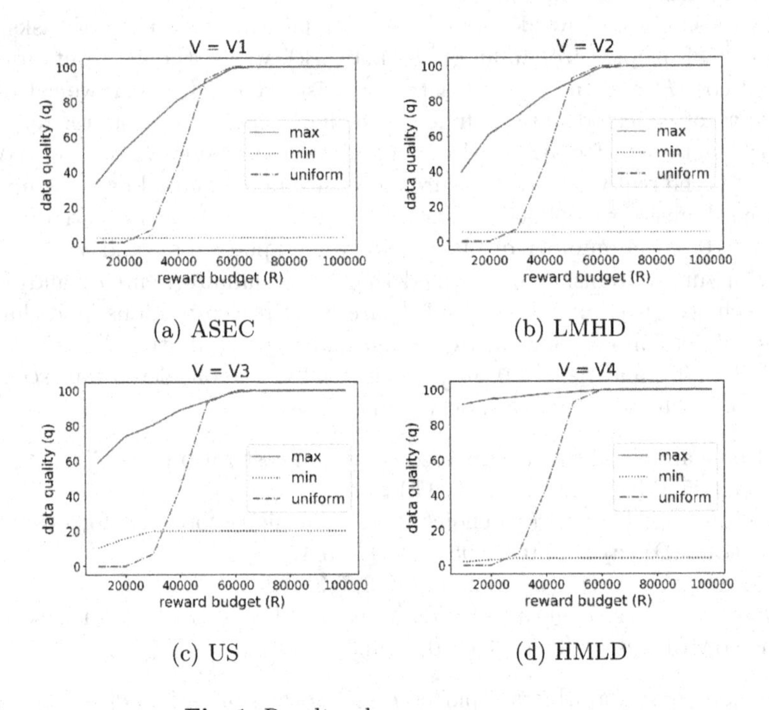

(a) ASEC (b) LMHD

(c) US (d) HMLD

Fig. 1. Results when $\boldsymbol{v} = \boldsymbol{v}_1, \boldsymbol{v}_2, \boldsymbol{v}_3, \boldsymbol{v}_4$

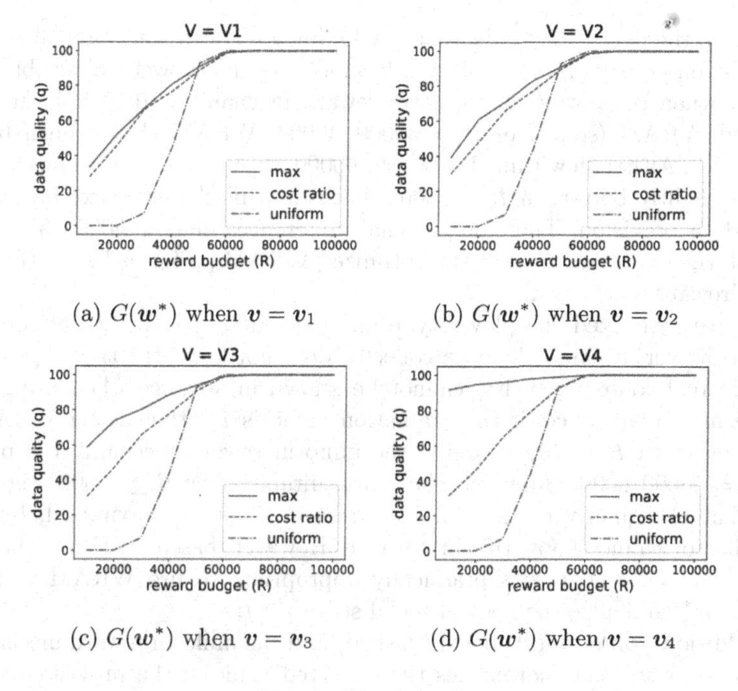

(a) $G(\boldsymbol{w}^*)$ when $\boldsymbol{v} = \boldsymbol{v}_1$

(b) $G(\boldsymbol{w}^*)$ when $\boldsymbol{v} = \boldsymbol{v}_2$

(c) $G(\boldsymbol{w}^*)$ when $\boldsymbol{v} = \boldsymbol{v}_3$

(d) $G(\boldsymbol{w}^*)$ when $\boldsymbol{v} = \boldsymbol{v}_4$

Fig. 2. $G(\boldsymbol{w}^*)$ results

5 Discussion and Conclusions

First, regarding the findings related to the uniform reward mechanism, shown as the dash-dot line in Fig. 1, because this is independent of the evaluation value $\boldsymbol{v}$, all the subfigures portray the same behavior, including a noteworthy increase of q_u from 0 to 100 in the range of $R = 20000$ to $R = 50000$. Therefore, almost no one responds to the questionnaire if the compensation budget R is smaller than the provision cost, but all respondents respond if it is sufficiently large. Subsequently, the results probably match the actual situation.

Second, specifically examining the WRAM results, the solid and dotted lines in Fig. 1 of $\max G(\boldsymbol{w})$ and $\min G(\boldsymbol{w})$ differ because the value of $G(\boldsymbol{w})$ depends on the evaluation value of the data collector $\boldsymbol{v}$ based on the definition of the quality of collected data $G(\boldsymbol{w})$. Moreover, a large difference exists between the values of $\max G(\boldsymbol{w})$ and $\min G(\boldsymbol{w})$ for any $\boldsymbol{v}, R$. This finding suggests that the quality of collected data $G(\boldsymbol{w})$ changes greatly depending on the setting of $\boldsymbol{w}$.

Third, comparing the uniform reward mechanism (dash-dot) and the worst WRAM (dotted) (Fig. 1a–1b), when $R = 10000, 20000$, the uniform reward mechanism is lower than the worst WRAM, which is attributable to the characteristics of the two mechanisms. In WRAM, the player decides whether or not to answer each questionnaire item. Therefore, even if R is small, there might be items that increase the utility by answering. In contrast, for the uniform reward mechanism, the player makes decisions by lumping all of the questionnaire items together. Consequently, if R is sufficiently small, then no answer will be obtained.

Next, when comparing the uniform reward mechanism (dash-dot) and the optimized WRAM (solid), in $R = 30000, 40000$, WRAM shows much better performance. Although within $R = 50000, 60000$, the uniform reward mechanism performs slightly better. In $R \geq 60000$, both data mechanisms exhibit similar quality data provision. Therefore, if small differences in $R = 50000, 60000$ are accepted, then one can say that the optimized WRAM performs better than the uniform reward mechanism.

Moreover, we fixed the player type and calculated the quality of collected data q. However, because the data collector does not know the true type of the player, in real contexts, such q cannot be known in advance when using each mechanism. For instance, if the evaluation value is $\boldsymbol{v}_2$, then using WRAM is convenient when $R \leq 40000$. Using the uniform reward mechanism is proper when $R = 50000, 60000$. Moreover, both are suitable when $R \geq 70000$. Figure 1b shows that the quality of the collected data q might be maximized, but the data collector cannot know this in advance. However, based on Fig. 2, the data collector can infer that it is practically appropriate to use WRAM with the weight of $\boldsymbol{w}^*$ to realize the desired social state $q \geq t$.

In addition, comparing WRAM (dashed) and the uniform reward mechanism (dash-dot) for $\boldsymbol{w}^*$ that normalizes the expected value of the provision cost in Fig. 2, when the compensation budget is $R \geq 70000$, both can get all answers. When it is $R = 50000, 60000$, the uniform reward mechanism performs slightly better. However, when the budget is $R \leq 40000$, the $\boldsymbol{w}^*$ of WRAM has a considerably larger G value.

Furthermore, let A be the area surrounded by the upper dashed graph and the lower dash-dot graph. Also, let B be the area of the upper dash-dot graph and the lower dashed graph. Thereby, A can be interpreted as the number of combinations of (R, t), where $q \geq t$ is realized when using WRAM with weight $\boldsymbol{w}^*$, but $q < t$ is realized when using the uniform reward mechanism.

However, B can be interpreted as the number of patterns for which only the uniform reward mechanism realizes the desired social state $q \geq t$. Subsequently, it is readily apparent that B is very small compared to A. Moreover, the same was true when the distributions of delivery cost F_1–F_5 were assumed to be uniform distributions over various different closed intervals.

By contrast, the comparison between the optimized WRAM (solid) and the WRAM with $\boldsymbol{w}^*$ weight (dashed) shows that WRAM with $\boldsymbol{w}^*$ is not optimized. However, as described above, the data collector cannot know the actual value of the weight $\boldsymbol{w}$ which optimizes WRAM.

Considering all of these findings together, it can be concluded that using WRAM with a ratio weight $\boldsymbol{w}^*$ equal to the expected value of the provision cost as a mechanism when conducting questionnaires is appropriate for data collectors seeking to improve the quality of the collected data.

In this paper, however, we assumed that all players do not lie and honestly provide accurate data they own. Because this assumption may not fit the actual contexts, considering the possibility of data submitters lying is our future work.

References

1. Agmon, N., Ahituv, N.: Assessing data reliability in an information system. J. Manag. Inf. Syst. **4**(2), 34–44 (1987). https://doi.org/10.1080/07421222.1987.11517792
2. Cao, G., Duan, Y., El Banna, A.: A dynamic capability view of marketing analytics: evidence from UK firms. Ind. Mark. Manag. **76**, 72–83 (2019). https://doi.org/10.1016/j.indmarman.2018.08.002. https://www.sciencedirect.com/science/article/pii/S0019850117306892
3. CrowdWorks Japan: Market price list of crowdworks that you want to check before making a job request (2022). https://crowdworks.jp/pages/guides/employer/pricing. Accessed 24 Dec 2022
4. Hazen, B.T., Boone, C.A., Ezell, J.D., Jones-Farmer, L.A.: Data quality for data science, predictive analytics, and big data in supply chain management: An introduction to the problem and suggestions for research and applications. Int. J. Prod. Econ. **154**, 72–80 (2014). https://doi.org/10.1016/j.ijpe.2014.04.018. https://www.sciencedirect.com/science/article/pii/S0925527314001339
5. Jones, M.: What we talk about when we talk about (big) data. J. Strateg. Inf. Syst. **28**(1), 3–16 (2019). https://doi.org/10.1016/j.jsis.2018.10.005. https://www.sciencedirect.com/science/article/pii/S0963868718302622
6. Keller, S.A., Shipp, S., Schroeder, A.: Does big data change the privacy landscape? A review of the issues. Annu. Rev. Stat. Appl. **3**(1), 161–180 (2016). https://doi.org/10.1146/annurev-statistics-041715-033453
7. King, W.R., Epstein, B.J.: Assessing information system value: an experimental study. Decis. Sci. **14**(1), 34–45 (1983)
8. Kyeong, N., Nam, K.: Mechanism design for data reliability improvement through network-based reasoning model. Expert Syst. Appl. **205** (2022). https://doi.org/10.1016/j.eswa.2022.117660
9. Mandal, P.: Data quality in statistical process control. Total Qual. Manag. Bus. Excell. **15**(1), 89–103 (2004). https://doi.org/10.1080/1478336032000149126
10. Manyika, J., et al.: Big data: the next frontier for innovation, competition, and productivity. McKinsey Global Institute (2011)
11. Miller, N., Resnick, P., Zeckhauser, R.: Eliciting informative feedback: the peer-prediction method. Manage. Sci. **51**(9), 1359–1373 (2005). https://doi.org/10.1287/mnsc.1050.0379
12. Miller, S.R., Bailey, B.P., Kirlik, A.: Exploring the utility of Bayesian truth serum for assessing design knowledge. Hum.-Comput. Interact. **29**(5–6), 487–515 (2014). https://doi.org/10.1080/07370024.2013.870393. https://www.scopus.com/inward/record.uri?eid=2-s2.0-84903120993&doi=10.1080%2f07370024.2013.870393&partnerID=40&md5=1c319d0d6d87f25159e9ad42335fc53b
13. Prelec, D.: A Bayesian truth serum for subjective data. Science **306**(5695), 462–466 (2004). https://doi.org/10.1126/science.1102081. https://www.scopus.com/inward/record.uri?eid=2-s2.0-5644221311&doi=10.1126%2fscience.1102081&partnerID=40&md5=ef029c9c1ed5459808da701d7eee6295, cited by: 264

14. Ramos-Lima, L., Maçada, A.C., Koufteros, X.: A model for information quality in the banking industry - the case of the public banks in Brazil, pp. 549–562 (2007)
15. Redman, T.: Data Quality: Management and Technology. Bantam professional books, Bantam Books (1992). https://books.google.co.jp/books?id=JCzMPAAACAAJ
16. Redman, T.: Data: an unfolding quality disaster. DM Rev. **14**(8), 21–23 (2004)
17. Schoenherr, T., Ellram, L.M., Tate, W.L.: A note on the use of survey research firms to enable empirical data collection. J. Bus. Logist. **36**(3), 288–300 (2015). https://doi.org/10.1111/jbl.12092. https://onlinelibrary.wiley.com/doi/abs/10.1111/jbl.12092
18. Sáenz, J., Ortiz de Guinea, A., Peñalba-Aguirrezabalaga, C.: Value creation through marketing data analytics: The distinct contribution of data analytics assets and capabilities to unit and firm performance. Inf. Manag. **59**(8), 103724 (2022). https://doi.org/10.1016/j.im.2022.103724. https://www.sciencedirect.com/science/article/pii/S0378720622001331
19. Strong, D., Lee, Y., Wang, R.: Data quality in context. Commun. ACM **40**, 103–110 (2002). https://doi.org/10.1145/253769.253804
20. Wang, R.Y., Strong, D.M.: Beyond accuracy: what data quality means to data consumers. J. Manag. Inf. Syst. **12**(4), 5–33 (1996). https://doi.org/10.1080/07421222.1996.11518099

Author Index

Y. Maemura et al. (Eds.): GDN 2023, LNBIP 478, p. 175, 2023.
https://doi.org/10.1007/978-3-031-33780-2

Author Index

Printed in the United States
by Baker & Taylor Publisher Services